Where this book is going

This book is about the greatest threat that *Homo sapiens* has ever faced—and what we might do about it. It is urgent in the same way that we would organize now to deflect a meteor strike due in 2035.

This analysis features a critique of the current climate message (gradual overheating, double down on emissions reduction), the extreme weather surge that occurred between 2000 and 2012, then a medical-school professor's justification for what makes it an emergency now, some design criteria for how to take the 50% excess of CO_2 out of circulation, and the proposal for how to get started with a "Manhattan Project 2.0".

It promotes a paradigm shift away from focusing on the next fractional-degree rise in temperature (where what to do about it is simply reducing CO_2 annual emissions) to a new focus on the extreme weather surge (where what to do is removing the excess CO_2 and an even faster Arctic cooling to reduce extreme weather). Here are some of the book's major illustrations to show you where the book is going.

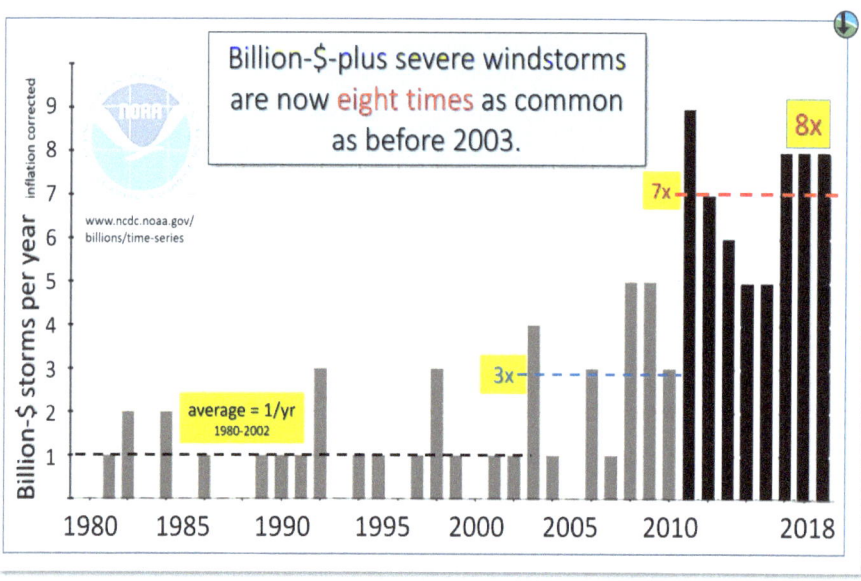

Figure 1. One of the five types of extreme weather that has surged since 2000.

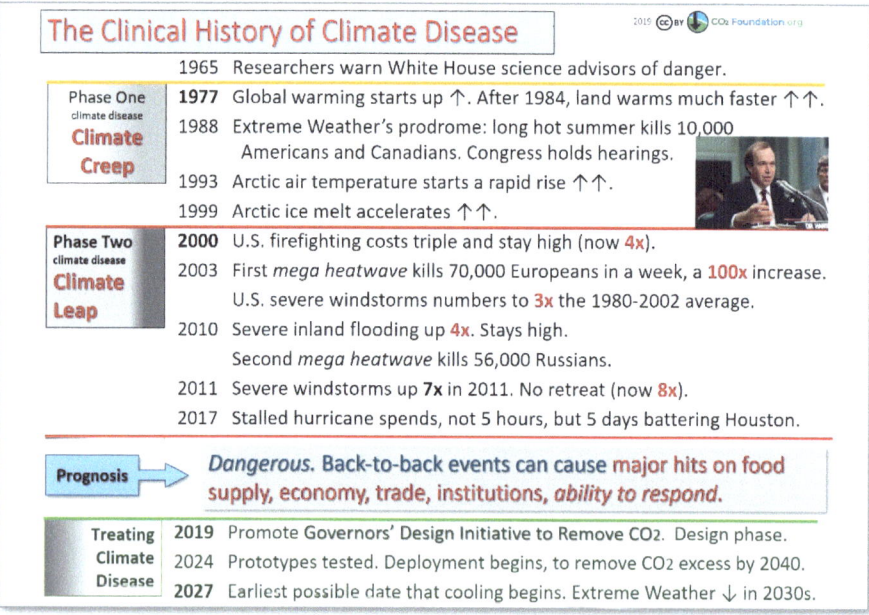

Figure 2. The "clinical history" of climate over 55 years.

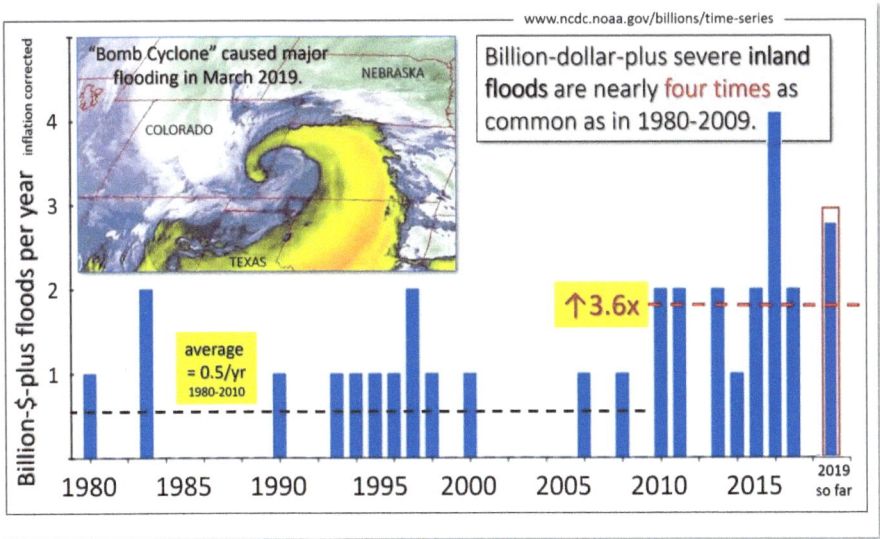

Figure 3. Billion-dollar floods surged in 2010.

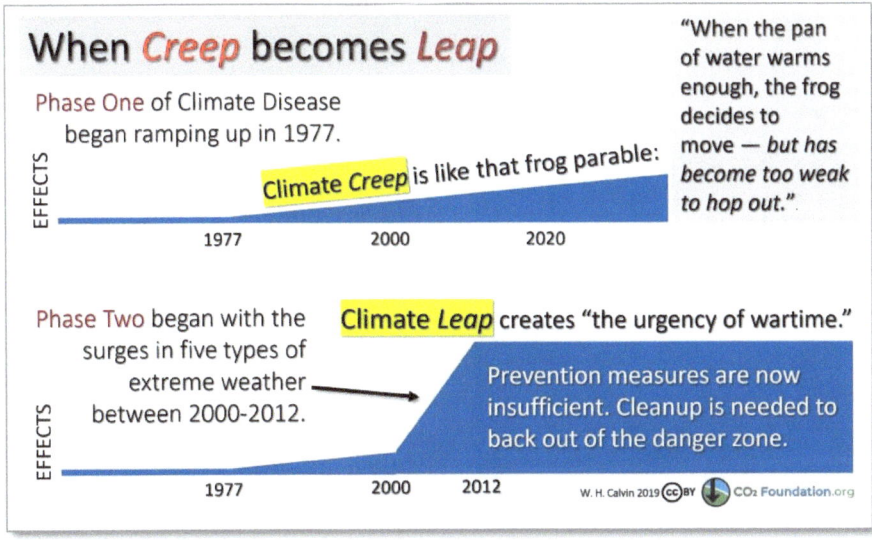

Figure 4. When climate creep became climate leap.

v

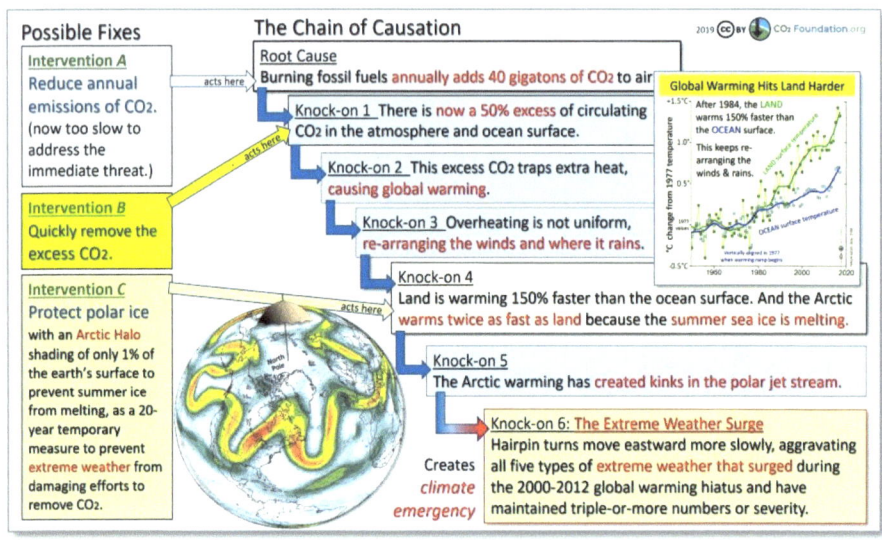

Figure 5. Knock-on effects in the climate cascade.

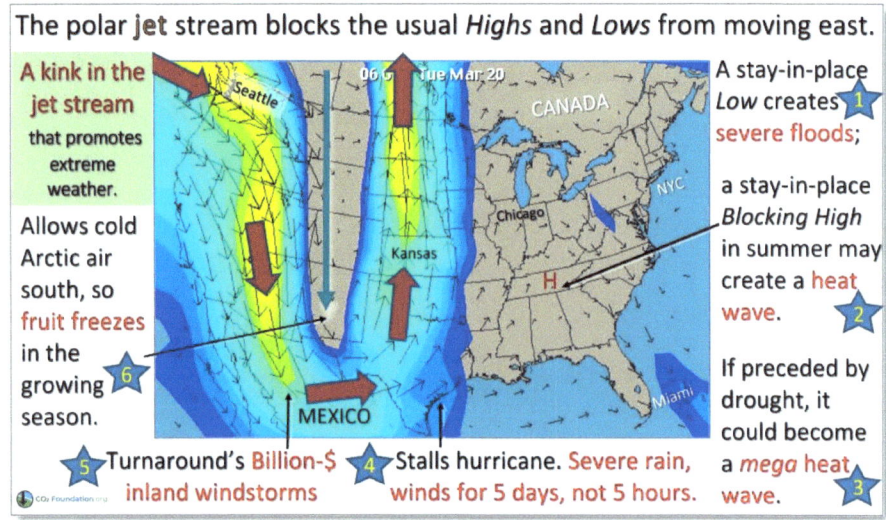

Figure 6. Hairpin turn illustrating the setup for six types of extreme weather. (They all did not happen on the same hairpin!)

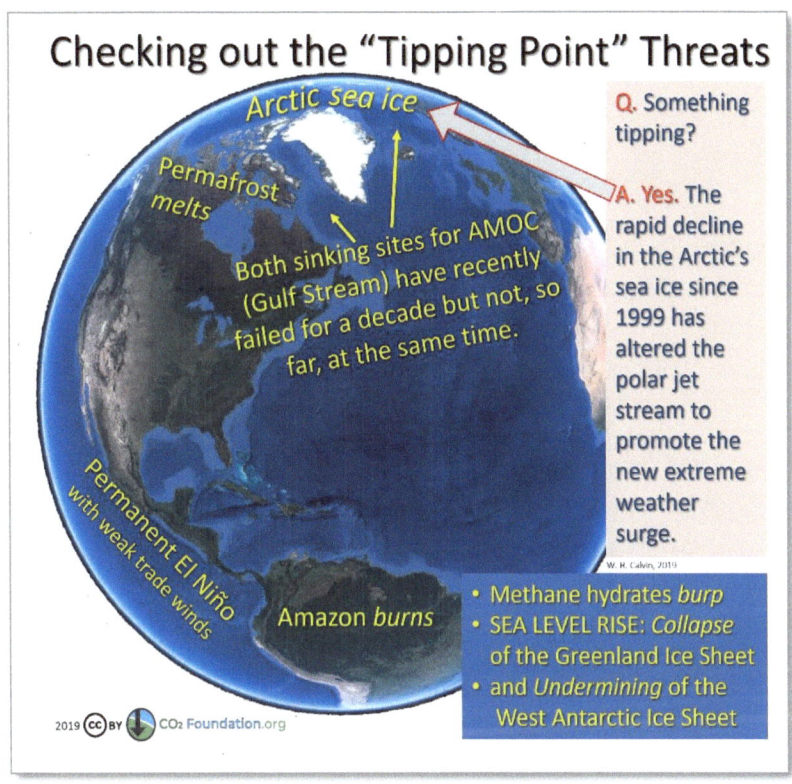

Figure 7. Timeline for Climate Repair

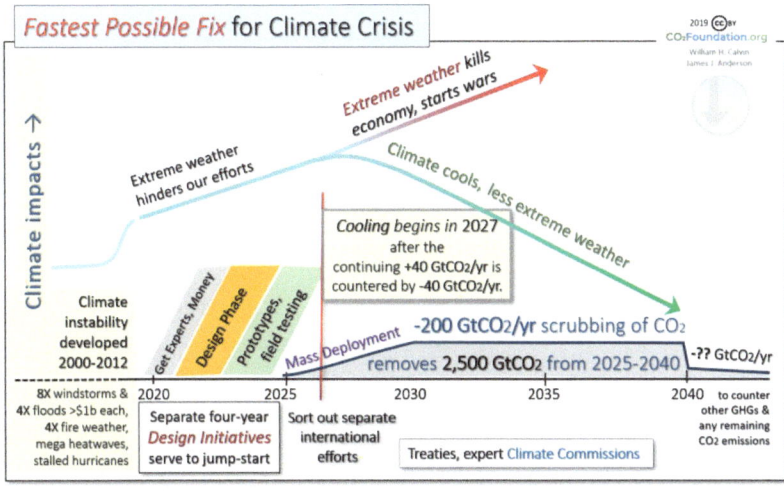

Figure 8. The usual eight tipping point candidates.

Previous Books by William H. Calvin

The Great Climate Leap (2012)

Global Fever: How to Treat Climate Change (2008)

Almost Us: Portraits of the Apes (2005)

A Brief History of the Mind: From Apes to Intellect and Beyond (2004)

A Brain for All Seasons: Human Evolution and Abrupt Climate Change (2002)

Lingua ex Machina: Reconciling Darwin and Chomsky with the Human Brain (2000), William H. Calvin and Derek Bickerton.

The Cerebral Code: Thinking a Thought in the Mosaics of the Mind (1996)

How Brains Think: Evolving Intelligence, Then and Now (1996)

Conversations with Neil's Brain: The Neural Nature of Thought and Language (1994), William H. Calvin and George A. Ojemann

How the Shaman Stole the Moon (1991)

The Ascent of Mind: Ice Age Climates and the Evolution of Intelligence (1990)

The Cerebral Symphony: Seashore Reflections on the Structure of Consciousness (1989)

The River That Flows Uphill: A Journey from the Big Bang to the Big Brain (1986. Revised edition 2010.)

The Throwing Madonna: Essays on the Brain (1983)

Inside the Brain: Mapping the Cortex, Exploring the Neuron (1980), William H. Calvin and George A. Ojemann

…and such major magazine articles as

"The great climate flip-flop" in *The Atlantic* (cover story, 1998)

"The emergence of intelligence" in *Scientific American* (1994)

"Memory's future" in *Psychology Today*, with Elizabeth Loftus (2001)

Websites: *faculty.washington.edu/wcalvin*, *WilliamCalvin.org*, *CO2Foundation.org*

EXTREME WEATHER
and what to do about it

William H. Calvin, Ph.D.

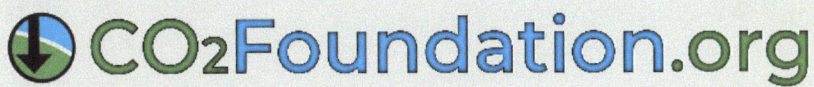

Copyright ©2019, 2020 by William H. Calvin, who has assigned it to the publisher, the nonprofit CO₂ Foundation of Seattle, Washington, USA.

The abridged softcover editions v0.9.1 have the same main text but, to lower printing costs, lack the endnotes and the Q&A of the unabridged editions. The 81 omitted pages are available as a free download from *CO2Foundation.org.*

ISBN 9781702353465 for mono abridged softcover.

ISBN 9781676025375 for the full-color abridged softcover.

The ebook/Kindle version is unabridged and in full color (if your device does not support color, use the web app).

Major Topic Headings

PREVIEW — iii
Where this book is going — iii
Grand Rounds for Climate Disease — 3
PREFACE — 5

SOME CLINICAL HISTORY OF CLIMATE DISEASE — 11

WHAT WE HAVE DONE ABOUT IT — 19
   Emissions Reduction — 20
   Adaptation — 21
   Even if it were working, "Use Less" is now far too slow — 21

WHAT'S NEW? — 23
1. Fire Weather Tripled — 24
   • Firefighting costs tripled in 2000 — 25
   • The Dalles wildfire in 2018 — 26
   • The 2018 'Camp' fire that destroyed Paradise — 28
   • San Francisco visibility in smoke — 29
2. Mega Heatwaves — 30
2a. *An American Mega?* How Heat Kills. — 35
3. "Those hurricanes that don't leave" — 41
   • The path of the 1991 "Perfect Storm" — 41
   • Hurricane Sandy in 2012 (a forced left turn) — 43
   • Hurricane Harvey (visited Houston for a week in 2017) — 45
   • Hurricane Dorian in 2019 — 46
4. Severe inland windstorms are up 8X — 47

4a. Familiar with the Derecho? ... 51
- The "Inland Hurricane" ... 54

5. Severe inland floods up 4X after 2009 ... 57
Five times more dengue virus circulating ... 58
Worse, but not tripled: long-lasting droughts ... 59
Worse, but not yet triple: "Crazy ice storms" ... 61

DIAGNOSIS (identifying causes) ... 63
Kinks in the jet stream ... 65
When the Jet Stream Wanders ... 66
    Detached Highs & Lows Spinning Down ... 69
    The Strange Year of 1988 ... 71
What's Behind Extreme Weather? ... 75

PROGNOSIS (if untreated) ... 81
From Creep to Leap ... 81
    Sea level rise and inundation extent ... 84
    Expansion of the tropics and the westerly winds ... 86
The Common Cause ... 89

WHAT TO DO ABOUT IT, *NOW* ... 92
    Chain of causation for extreme weather surges ... 93
An appraisal of useful climate actions ... 97
    Small symbolic steps ... 98
    Doubling the forests ... 101
    The time scale of the climate emergency ... 102
When to do? ... 103
A paradigm shift to repair climate ... 106
A Course of Climate Action ... 107
    Climate engineering and two simplified examples ... 108

The ocean's carbon cycle: The in's and out's of CO2 110

THE TREATMENT PLAN — 113

The 2020 Manhattan Project — 113
 Treatment plan for climate disease 114
 Timeline for Defeating Extreme Weather 115
Governors' Design Initiative to Repair Climate — 121
The Wrap — 123

Q&A — 125

 "How many big emergency projects are we talking about?" 125
 "Sea level rise? Has that train left the station?" 127
 "Say more about that chain of causes." 128

The next 80 printed pages have been omitted from the abridged softcover editions to lower costs. They are included in the ebook edition and are also a free download from *CO2Foundation.org*.

 "What is *mitigation*? Is it like adaptation?" 128
 "What do you say to the governor's environmental advisor?" 130
 "People keep trying to stop geoengineering?" 131
 "Say more about those design criteria, please." 132
 "So, do the governors pick the best design?" 133
 "How fast is *fast*?" 135
 "Why was the abrupt climate change ignored for so long?" 136
 "Why have our climate actions been so slow?" 137
 "Why store carbon in the ocean depths?" 140
 The life cycle 142
 "Why won't ocean fertilization work by itself?" 143
 Mimicking ocean downwelling and upwelling 144
 "Won't you cause fish kills with such blooms?" 145
 "Where do we put the plankton farms?" 147

Talking climate 149

"Why the Continental Shelf?" 150

"Why a Climate Emergency?" 151

"Were climate science reports too conservative?" 154

"So, emissions are like packs-per-day?" 155

"So where did the surges come from?" 157

"What about that vicious cycle, where cooling creates more emissions?" 159

"But what about methane?" 160

"How could we double forests quickly?" 161

"Is the Amazon going to dry up?" 162

"Does the planet pay any attention to its average temperature?" . 165

Regional differences in warming rate 166

Careful what you lump together 167

"It's not the heat. It's the humidity." 168

"What's a derecho, again?" 170

Hammered by Downbursts 171

Caught Unprepared 172

"What about derechos outside the U.S.?" 173

"What's with Arctic Amplification?" 175

"That Arctic Halo is temporary?" 176

"How does one create an Arctic Halo?" 179

"Any runways that are close enough?" 179

"Explain yourself. How did you get from brains to climate change?" 183

Scientific Sources   193

Readings 193

Endnotes 195

With thanks to the directors of
the CO_2 Foundation,
who assisted in many ways:

James J. Anderson
Katherine Graubard
Raz E. Mason
Sally McDonough

"A scientist has to be neutral in his search for the truth, but he cannot be neutral as to the use of that truth when found.

If you know more than other people, you have more responsibility, rather than less."

—C. P. Snow

"Hear this, young men and women everywhere, and proclaim it far and wide. The earth is yours and the fullness thereof. Be kind, but be fierce. You are needed now more than ever before. Take up the mantle of change. For this is your time."

—Winston Churchill

Grand Rounds for Climate Disease

This book's topics are ordered much as a medical case is presented at Grand Rounds (the weekly detective-story conference at any medical school):

- A patient's clinical history (here, the history of normal climate fluctuations, then of the global overheating).
- The current signs and symptoms (the new extreme weather), together with lab results.
- Possible diagnoses (jet stream kinks, caused by the Arctic's acceleration of global overheating from fossil fuel CO_2), including the chain of causation.
- The prognosis (establishing how urgent treatment is).
- Potential treatments (emissions reduction, reflecting sunlight, CO_2 removal) and how to avoid them harming the patient via side effects.
- As the finale, the proposed treatment plan (here, the *Governors' Design Initiative to Repair Climate*).
- Q-and-A after the presentation proper, expanding on selected aspects.
- Grand Rounds does not always focus on a single patient. Sometimes it is about a new way of looking at things, certainly the case for my occasional presentations over the years at neurosurgery, neurology, and psychiatry & behavioral sciences grand rounds.

This "Grand Rounds" will be about identifying Phase Two of climate disease, distinguishing what makes it so different from Phase One, and what makes it so threatening.

The disease progression recently went from Climate Creep to Climate Leap, even though the global overheating paused for a dozen years—all while the CO_2 accumulation continued to creep up. Strange, but phase transitions and complex systems do such things. Such a paradox shows why we need a rethink.

Figure 9. Home, seen rising over the moon from a speeding satellite.
Credit: NASA

PREFACE

I am going to try a different way of presenting civilization's biggest challenge. Let me illustrate how this emerging climate crisis would be viewed by those in the medical community who are used to dealing with fuzzy categories and their closing windows of opportunity for effective action.

I hope to provide a better intellectual toolkit for facing up to the climate crisis, largely borrowed from what we teach medical students about dealing with emergencies.

There are now greater concerns than when the global overheating ramp hits 1.5°C or 2.0°C. Focus your mind for a moment on the traditional bathroom light switch, where a gradual increase in finger pressure suddenly triggers a flip and a click, flooding the space with painfully bright light.

Climate can be bistable as well. We keep assuming the overheating is like pushing a dimmer switch, where results are linearly proportional to the push. For five types of extreme weather, something flipped between 2000 and 2012. It was *regime change*. It changes everything—it even makes statistics of the past worthless in some cases.

That's what this little book is about, not the better-known gradual view that others have adequately described—and for fifty years.

...

"Think Fast. And then Think Again" is a motto for the emergency medicine physicians. Get moving, but keep revisiting the patient's diagnosis, in case something additional has crept in—say, internal bleeding or shock.

During quiet times in the office, some physicians get around to asking themselves if their standard treatments really work. For climate, we too must ask: *Has our fifty years of emissions reduction been effective? Can they ever?*

Despite occasional near-success stories, such as the way that California has held emissions per person constant since 1980, the annual world-wide bump up in carbon dioxide has increased 50 percent since the turn of the 21st century. This is not progress.

And in the future? About a third of annual emissions now come from the less developed countries, soon to be in great need of overnight air conditioning to survive heatwaves. They will then burn their local fossil fuels to generate electricity to run the extra A/C units. But it's a global common because of air mixing: their CO_2 doesn't stay local. Just as ours did not.

The continued framing of climate action as an **emissions reduction** task (similar to a heavy smoker "cutting back to one pack a day"), with no additional discussion of backing out of the danger zone for extreme weather, will take us straight into "too little, too late" and the massive social consequences of hopelessness. We do not want to go there.

Prevention and treatment often demand different approaches, but that is often not reflected in major international scientific reports about our climate problem. Civic organizations supporting climate action usually uncritically echo them, focusing solely on rallying their troops to use less.

Most reason, in effect, "Emissions caused the problem, reducing them ought to fix it." However true for smog cleanup in the 1970s, CO_2 is not cleaned up by nature as fast as visible air pollution is (a thousand years vs. two weeks). No one ever mentions that.

Today, the continuing emphasis on "use less" is like treating a painful tooth solely with reduced candy consumption. While emissions reduction was the obvious strategy for CO_2 fifty years ago, it is a preventative measure (like reducing smoking), not a fix once a disease (like lung cancer) develops.

Things have changed, but our strategy has not. Start asking why.

. . .

Are our leaders successfully warning us of imminent danger? The scientific leaders try, but people may compare our latest +1.5°C scientific warning to the occasional hotter summer or to a mild fever. They opine, "That's not much. What's the big deal? Just wait and it will go away," not realizing that nature takes so long to do the CO_2 cleanup job. We must do the cleanup ourselves and the new extreme weather says we must do the cleanup very quickly.

Fifty years of trying to warn people using the creep of small numbers is enough. It is time to lead with other indicators of trouble. I'd suggest leading with the *maintained* surge in extreme weather. Those escalation numbers are big. And they are recent. Many have felt them; they do not have to imagine a warmer future.

Current extreme weather threats could crash the economy and leave us too battered to get our act together for effective action.

This alone demands a change in strategy. We now need stronger climate medicine in the form of a CO_2 cleanup.

The window of opportunity for backing out of the danger zone may be as short as the next ten to twenty years. For even big CO_2 scrubbing projects, it will take at least eight years to get started cooling, and we don't know how fast climate troubles will decline as the CO_2 comes down. This makes an immediate start even more urgent.

It might take four years to get the U.S. Congress up to speed. *That's four years that we don't have,* not anymore. Replacing reluctant legislators is now too slow; we must help them open their minds instead. Yet we also need to get moving on climate repair this year.

. . .

In the U.S., we worry it will take another decade to get our government started on a big climate repair project. That, I suspect, will be too little, too late.

It will lead to widespread despair. For the massive social consequences of hopelessness, one hazard we face are mass suicide-homicide schemes. Look up[1] *Jonestown* in Guyana in 1978 (a cult from Indiana), *Heaven's Gate* in southern California in 1997, and the cults which aspire to become mass murderers—say, the wealthy Japanese-Russian *Aum Shinrikyo* cult's deadly neurotoxin attack at the Tokyo underground's government center station in 1995. Via blaming major nuclear powers for the killings, "Aum" aspired to trigger a nuclear winter.

The Aum faithful[2] would, of course, survive to inherit the earth. The gullible included some technically well-educated people, including physicians and the chemical engineers who created the sarin gas.

This isn't fantasy fiction but rather some recent history. Surviving the next heat wave might be the least of our immediate problems.

. . .

There is a possible workaround to climate action's slow road ahead, one that would provide hope as well. The CO_2 Foundation is reaching out to a few state governors to immediately set up and run a nonprofit design project, where a group of experts work together for a few years to design and prototype a much bigger Carbon Sink undertaking, one that removes the excess CO_2 from circulation. The experts would aim for prototypes and field trials done in only four years, like the Manhattan Project did between 1941 and 1945.

The *Governors' Design Initiative for Climate Repair* would utilize a finance committee of a dozen tech billionaires, already familiar with what it takes to complete big design-and-prototype projects quickly. The governors would run the project, likely via hiring an experienced serial entrepreneur to act as general manager.

News about the project would help build momentum for legislative action authorizing mass deployment. Without such a project to provide an expert stamp of approval, legislators may be unlikely to act quickly. The designs would be available to all countries, allowing faster roll-out than any market-driven solution could achieve.

Enough scrubbing capacity to remove the excess CO_2 by 2040 is the goal, yet even that would not get cooling started for at least eight years (the ramp-up needs to first counter the continuing emissions).

A lot can happen in the meantime, but our situation is not hopeless, as exaggerated headlines are starting to suggest. Our situation is not that bad. There are effective actions we can still take, if we hurry.

WHC

N.B. I am told that I need a brief bio here (even though there is a longer one at the end of the book).

I am neither a climate scientist nor a science journalist; rather, I am a basic scientist (Ph.D. in Physiology & Biophysics) in the neurosciences, coming from a physics background. Decades of contact have also acquainted me with the thinking habits of academic psychologists, archaeologists, physical anthropologists and, since 1983, the climate scientists (mostly the atmospheric scientists, geophysicists, and physical oceanographers) who have invited me along to their summer retreats for the last fifteen years, a great help when writing my three previous books on climate.

I also have more than a half-century of hanging around a medical school. I was appointed to the neurosurgery faculty at the University of Washington before I finished my Ph.D. in 1966, was appointed to the biology faculty in 1985, and then moved to psychiatry & behavioral sciences in 1992. In 1998, I was elected a fellow of the Association for Psychological Science.

I gradually absorbed, mostly from daily coffee-room discussions with the neurosurgeons, many of the thinking habits of the real pros at emergency management. Dealing with emergent situations requires a different mindset. Neither climate scientists nor the relevant policy wonks seem to have it yet.

I am now the president of the nonprofit CO_2 Foundation in Seattle. It's a sign of the times that I also need to add that I have never worked for the fossil fuel industry, nor do I invest in fossil fuels except via the big pension funds where I have no choice.

Figure 10. Annual emissions lead to CO2 accumulation overhead, which in turn leads to global warming after 1976. But the warming is uneven after 1984 (seen below with five-year smoothing), with the continents warming much faster than the sea surface. Largely because of ice-albedo feedback, the Arctic warms twice as fast as the adjacent continents. That has disrupted the polar jet stream to cause much more extreme weather globally.

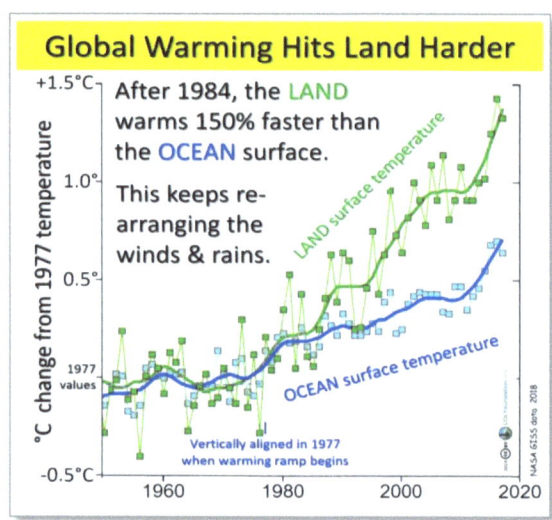

SOME CLINICAL HISTORY OF CLIMATE DISEASE

Concern about human actions changing the climate goes back to the German scientist Alexander von Humboldt in 1814. When he wasn't traveling, he mostly lived in Paris. Humboldt did a lot of public speaking about his scientific travels and was said, at the time, to be the second-most famous person in the world (after Napoleon). He was the first famous environmentalist.

His climate concerns were expressed even before the heat-trapping properties of CO_2 were discovered in the 1820s by Jean Baptiste Joseph Fourier, better known for an infinite expansion in math, the Fourier Series that he developed for calculating the heat flows.

Research progress on global overheating was spotty after 1824, with long gaps between major developments (in 1859, 1896, and 1938). After 1955, scientific interest in the problem intensified. Monthly measurements of CO_2 began atop Mauna Loa in Hawaii in 1958, as far removed from industrial sources as possible, and 1979 marks the year when satellites began to carry the right scientific instruments.

That's also about when the tree-ring-style history of climate change was augmented, as the glaciologists began drilling ice cores in the Greenland ice sheet to discover a deeper history of temperature, precipitation, and high winds—and sometimes able to distinguish one year from the next. Then the climate scientists managed to extract ancient CO_2 from the bubbles in the ice, extending the record back to 800,000 years ago.

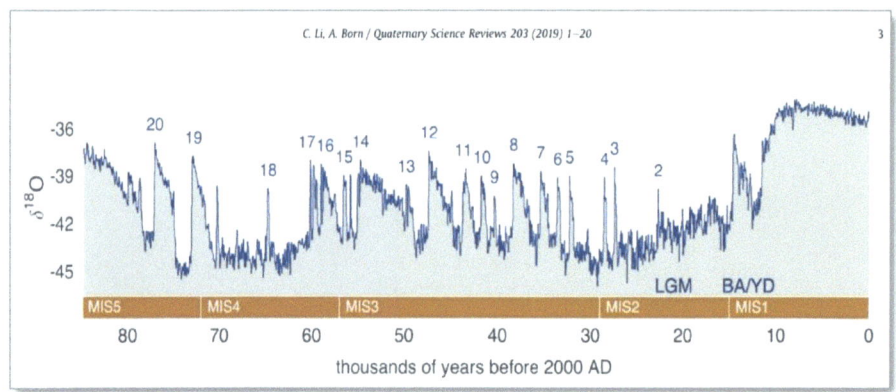

Figure 11. The North GRIP core from Greenland Summit. Up is warmer, down is cooler. The numbers are the D-O events, looking back from sudden warming event at 15,000 years ago. The vertical range in the oxygen isotope ratio covers about 6°C in the air above the North Atlantic Ocean.

This ice core is from highest point in North Greenland. Those jumps *up* only took about five years, and that's why we know that we are dealing with a bistable system.

. . .

A familiar term for heat-trapping is the "Greenhouse Effect"—but those are scare quotes because greenhouse is an inadequate analogy. The glass roof of a greenhouse keeps hot air from rising further, as it usually does, and so cooler air does not arrive from the sides to fill in. The so-called 'Greenhouse gases' do *not* work in this manner; there is no barrier. The CO_2 molecules in the air overhead absorb energy from the outbound infrared radiation from the earth's surface. So stimulated, they then emit their own infrared radiation in all directions; half of those directions are back down to the surface. On Venus, where the atmosphere is 96% CO_2, this effect makes the surface extremely hot (462°C/863°F), much more than Mercury, which is closer to the Sun.

Between 1950 and 1977, "Is the earth really warming?" became an important scientific question[3]. It should have been warming, according to the theoretical prediction from the rising CO_2—*but it wasn't*.

Since 1977, it has usually been warming. And so, "Is the earth *really* warming?" has been an answered scientific question[4]. It's a fact, verified in many different ways—those who don't trust thermometers can rely instead on ice melting, earlier budding of trees, later autumn freezes, the expansion of the tropics, and a dozen other indicators.

Facts are the only things that one can truly rely upon in this world; to survive, our beliefs must be adjusted to fit the facts. In science, we say a theory is "constrained by the facts."

The oceanographer Wally Broecker got it right in 1975, before the overheating ramp began in 1977. He said

> *Over the past 30 years, the warming trend due to CO2 has been more than countered by a natural cooling. This compensation cannot long continue, both because of the rapid growth of the CO2 effect and because the natural cooling will almost certainly soon bottom out. We may be in for a climatic surprise.*
>
> —WALLACE BROECKER *in Science 1975*

(Were Wally still alive, he would likely change that 'natural' in front of 'cooling'. There are anthropogenic candidates now.)

It is as if global cooling capacity maxed out[5] in 1977. Before 1977, some climate cooling mechanism had apparently opposed the warming from more and more CO2 in the air. Broecker figured out that it was not man-made dust aloft. (Just as in a medical diagnosis, one spends a lot of time eliminating the less-obvious possibilities.)

Perhaps there was more cloud cover, perhaps it was other air pollution reflecting more sunlight back out into space. The issue isn't settled yet, largely because that was back before good satellite coverage for measuring such things began in 1979, allowing a database record.

. . .

Today, when you hear global warming questioned, it often reflects the success of paid propaganda. The fossil-fuel industry didn't even invent this strategy to maintain profits in the face of bad news. The *doubt-the-scientists* tactic was borrowed from the deeply cynical deny-the-medical-science playbook of earlier campaigns to maintain asbestos and tobacco profits in the face of public health concerns.

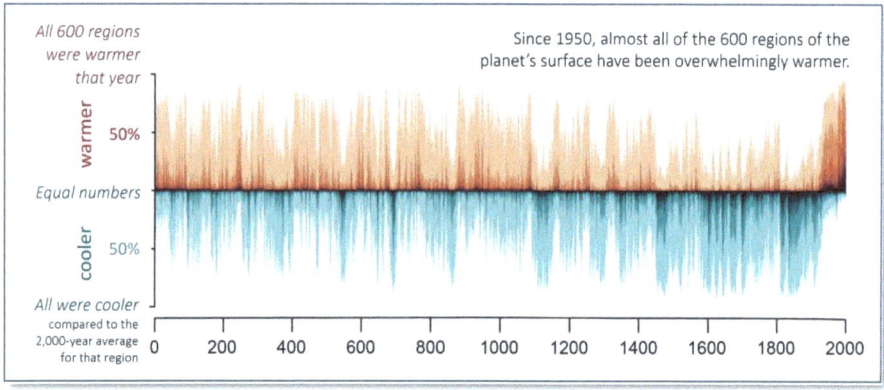

Figure 12. The global overheating trend, where each year's warmer/cooler is judged from the local average there over 2,000 years.

No one should have been confused on the second and third time around, but many were. Apparently, legislators confused by "Blowing smoke" tactics will postpone action for decades. Such unethical propaganda bought the tobacco industry a half-century of extra profits. Deny-the-science has been effective in postponing climate action for almost as long.

. . .

Consider: There have been two mega heatwaves so far; the 2003 mega in Europe killed 70,000 people; the 2010 mega in Russia killed 56,000. The U.S. is, of course, *not* preparing for a North American mega—such talk, someone has reasoned, might depress real-estate values.

It's not just those homeowners who are thinking of moving to a cooler place, hoping to whitewash a threat until they can sell at current prices. Dropping prices might even make banks unstable from too many 'underwater' loans. When banks speak, Congress listens.

But this roadblock from *self-interested denial* is now endangering the rest of us. We cannot afford it. At some point, the economy and perhaps even civilization itself will break, allowing revolutions and civil wars to distract us from organizing repairs to the climate. (Save your revolutions, if any, until after climate is repaired; history shows it takes decades to reorganize collapsed government services after winning.)

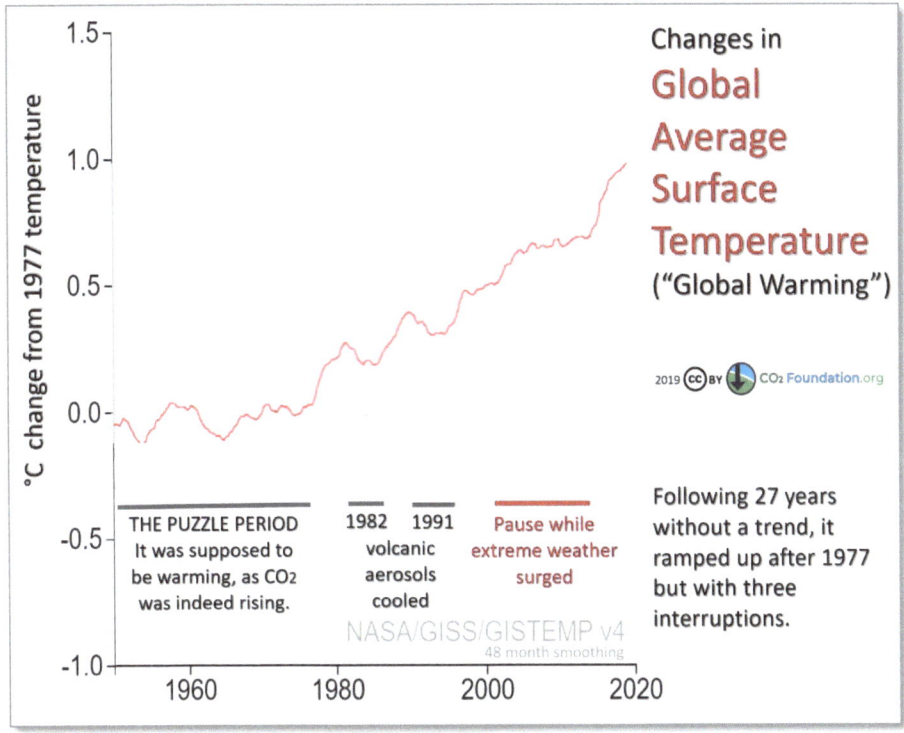

Figure 13. Global average surface temperature 1950-2019. This curve is a four-year smoothing of annual temperatures.

. . .

Why should the states most affected by climate change be the most anxious to deny what climate science says about the future?

The old answer to this paradox, back when fossil fueler propaganda seemed the obvious answer, was that they have longer distances to drive and above-average energy consumption for heating and cooling, making them price sensitive. In those states, single-family homes are the standard. But an urban apartment only loses and gains heat from one or two outside walls; heat transfer to neighboring apartments is radiated back in almost equal amounts. Houses lose/gain heat from all four sides, tops and bottoms, making per-capita energy consumption much higher. But since few people know that, it seemed an unlikely motivation.

These days, I suspect the aforementioned declining land value, plus the mortgage trap. That's enough to make bankers worry, and they are as influential in the state legislatures as are the providers of fossil fuels and the railroads that haul coal cross-country (also big contributors to the anti-science propaganda funds).

What parts of the U.S. are currently experiencing the worst of coastal flooding? Or humidity-augmented heat waves? Or the new extreme weather? Here are a few dozen recent examples of big extreme weather episodes in the U.S. Note that, except for fires and drought, most are in the South, Midwest, and the Great Plains to the east of the Rocky Mountains.

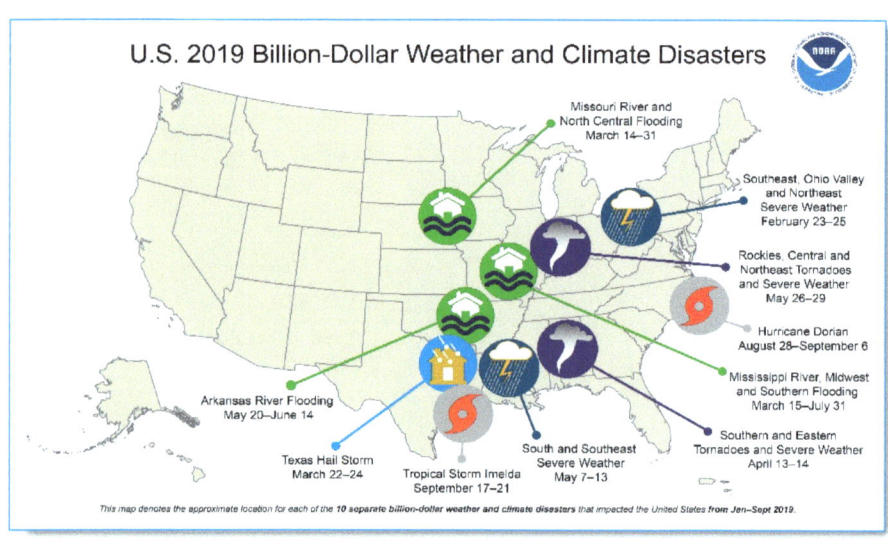

Figure 14. Map locating 2019 billion-$ extreme weather disasters.

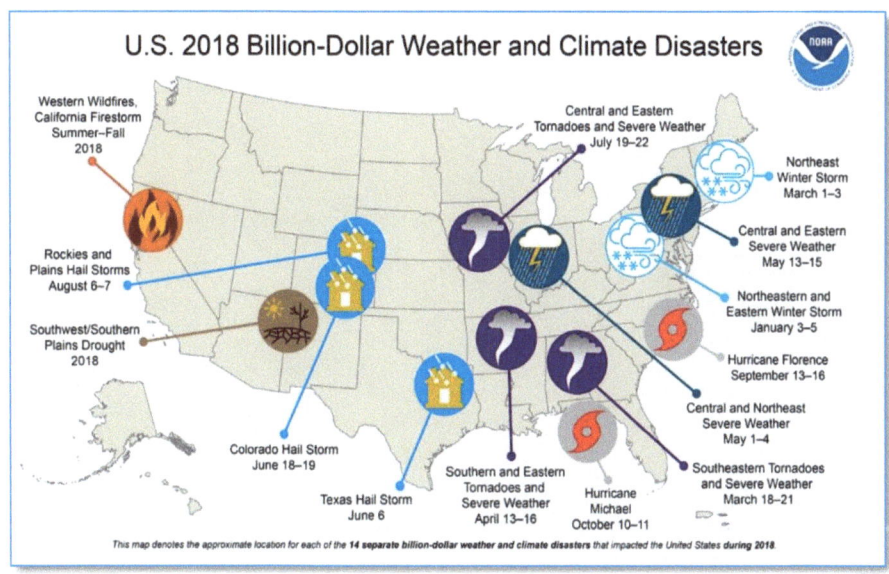

Figure 15. Map locating 2018 billion-$ extreme weather disasters.

The most-affected states ought to be the ones ringing the alarms and demanding federal climate action—but they instead elect to federal office most of the climate deniers and government downsizers, while collecting most of the federal crop insurance payouts, and much federal disaster

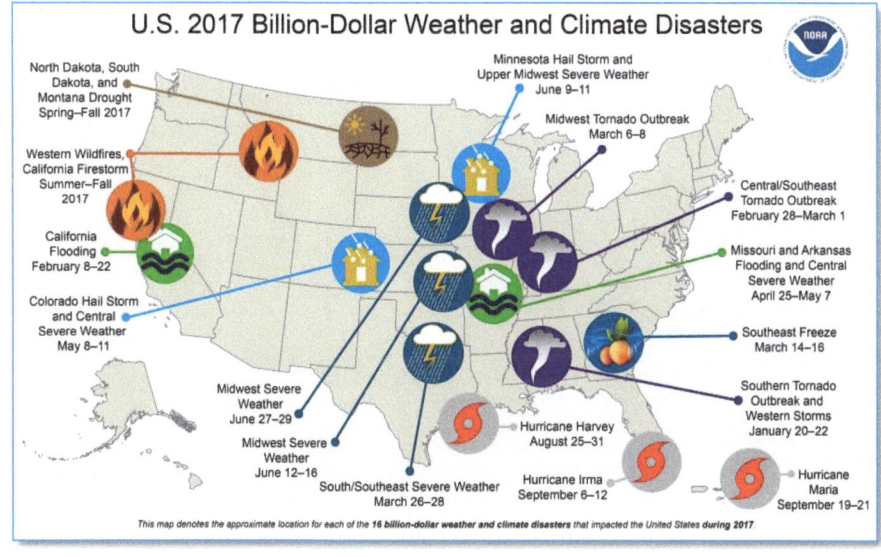

Figure 16. Map locating 2017 billion-$ extreme weather disasters.

17

aid. It makes no sense that they have been persuaded to be climate deniers, as it goes against their economic interests. Why?

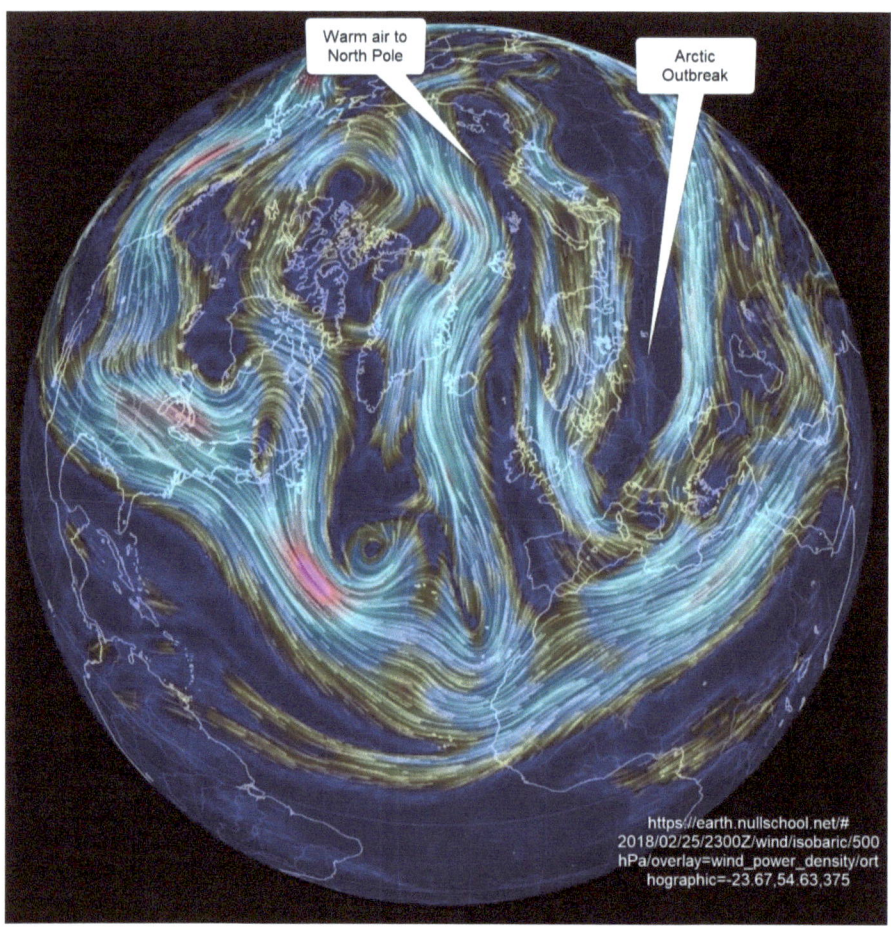

Figure 17. The polar jet stream disorganization on 25 February 2018 that led to the North Pole thawing in 24/7 winter darkness. We also see an Arctic outbreak reaching the Mediterranean, with an adjacent path for warm air from Ireland to reach the North Pole. Both extremes are a result of the loopiness of the polar jet stream.

WHAT WE HAVE DONE ABOUT IT

Figure 18. The three partial fixes for climate disease.

Emissions Reduction

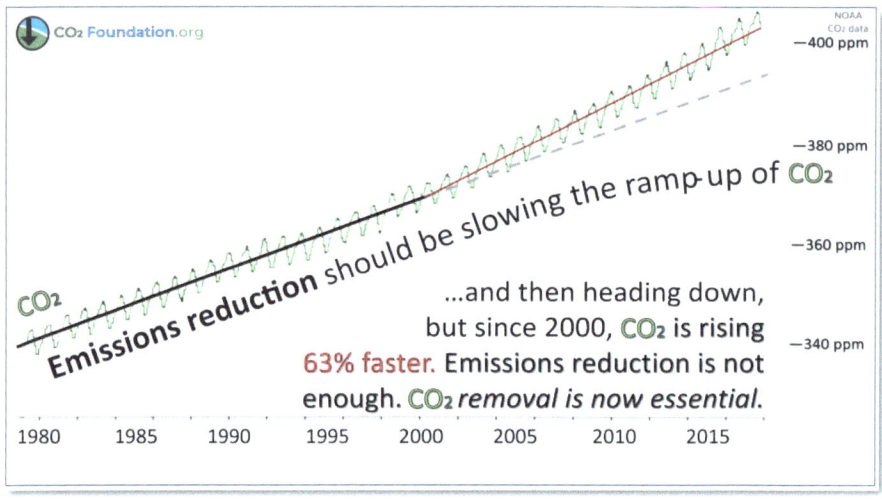

Figure 19. CO2 historic record. **Excess CO2** is the amount in excess of the concentration about 1800, 280 ppm. The excess is now 50%.

Examples of emissions reduction go under such names as clean energy, zero emissions, electric cars, carbon tax, etc.

There are some logical fallacies at work because even our leaders use 'emissions' as an abbreviation for *annual* emissions. That leads to confusion. Emissions reduction is not aimed at reducing the existing CO_2 *accumulation* (what causes the overheating) but at the next *annual addition to the accumulation*.

The efficacy of emissions reduction has been oversold; a headline containing "climate solution" is usually a very exaggerated claim about a very small effect that will arrive too late. A great deal of "Me too" has been going on via press release headlines that are uncritically printed.

Besides emissions reduction (a form of mitigation), there is talk of adaptation.

Adaptation

Climate adaptation includes measures needed to live in a climate-changed world—the common example is more air conditioning—but it does not include the preventative and slowing measures such as "Use Less" mitigation does.

Constructing new apartment buildings underground in 55°F/13°C deeper soil, as was done in previous centuries for vegetable cellars or ice houses, would be an effective adaptation in some places. 'Heat' (short for heat-sink) pumps are a type of air conditioner that cools surface buildings by circulating radiator fluid down to where it can be re-cooled to 55°F/13°C; that cuts down on the electrical power needed for air conditioning via the standard refrigerator compressor methods.[6]

Re-location of many coastal populations is required between now and 2050—even if we manage to cool air temperature by 2040, as the ocean surface will be slower to cool and shrink. The other classic creep in air circulation, expansion of the tropics, may respond somewhat more quickly but the fluctuations in where the winds blow, and rain falls, will create serious problems for agriculture in the meantime—which means a lot of new infrastructure to move water around.

So, even if there were no battering from extreme weather, governments still need to quickly start relocating coastal populations.

Even if it were working, "Use Less" is now far too slow

Recent estimates show that 26 percent of all the carbon released as CO_2—from fossil fuel burning, cement manufacture, and land-use changes over the decade 2002 to 2011—was absorbed by the oceans. About 28 percent went to plants and roughly 46 percent to increasing the accumulation in the atmosphere that overheats us some more. *About half of what goes up, stays up.*

Figure 19 showed the annual growth in atmospheric CO_2 since 1979, what was left in the air once the oceans and plants had removed what they could from annual emissions. The sawtooth peak is the Northern Hemisphere winter when leaves rot to generate CO_2 and many trees are dormant, no longer removing CO_2. The annual swing down is from plant growth in northern spring and summer, temporarily removing some CO_2 from the air.

You would think that the reversed seasons of the Southern Hemisphere would smooth out this sawtooth, but there is much less ice-free land surface below the equator and thus much less land-based greenery. Add up Australia and New Zealand, the half-as-wide part of Africa south of the Congo, and South America to the south of the mouth of the Amazon River. Compare that to the extensive land surface north of the equator, especially at higher latitudes such as 60° to 70° with lots of summer sunshine in the nighttime hours.

Estimating annual emissions depends on how well each of 195 countries report their contribution. Better to measure the resulting CO_2 itself, currently a 3 ppm bump up each year, as it is what adds to the *accumulation* that causes the overheating and climate shifts. In the 21st century, the annual bump-up in CO_2 *increased* by 63%.

Not exactly progress, is it? Let me repeat: in relying exclusively on an emissions reduction strategy, *we have been "betting the farm"* on something that *will not do the job that now needs doing.*

Speak firmly to our leaders; make sure they understand the new dimensions of our climate crisis, what makes it a *climate emergency*. Say that you expect them to lead, not lag—and certainly not obstruct.

If climate's demolition derby might destroy civilization in the next few decades, emissions reduction should not be where you direct most of your efforts.

WHAT'S NEW?

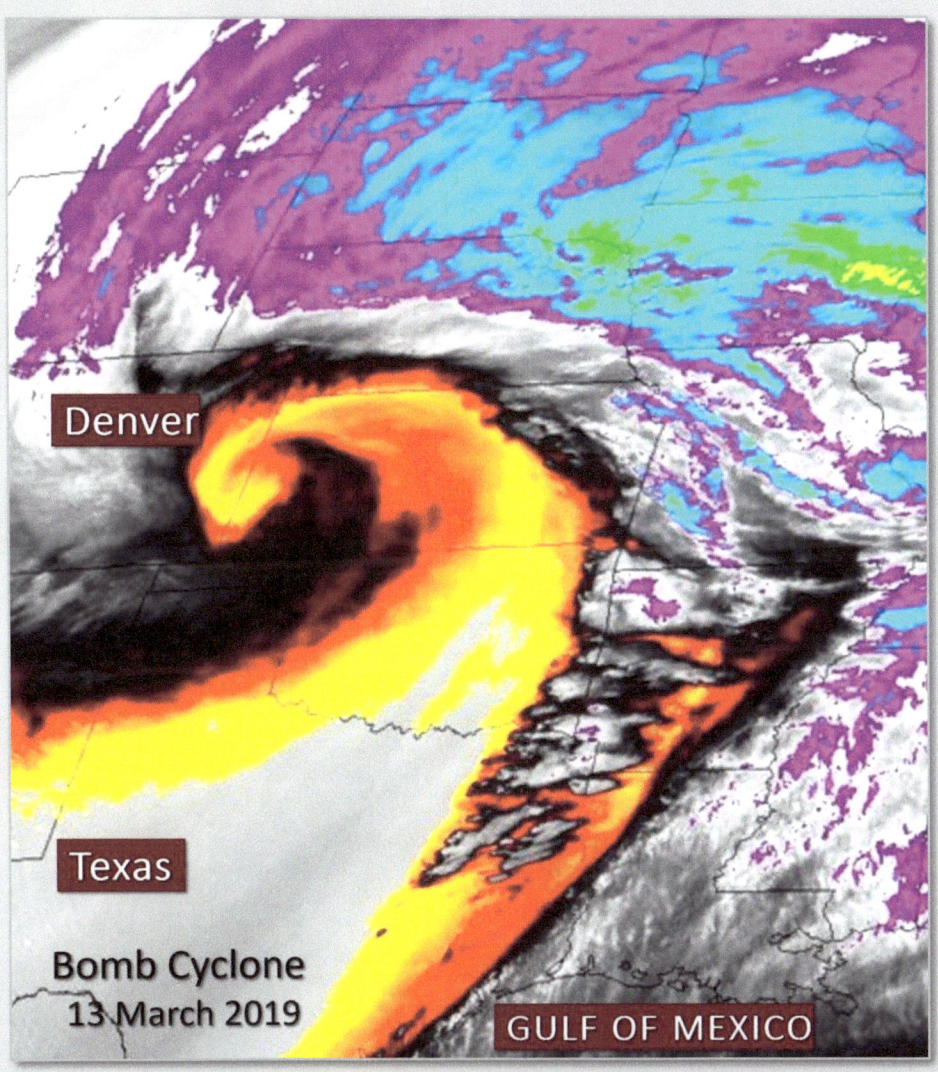

Figure 20. One of the three billion-dollar-plus floods in the U.S. in 2019 was caused by this "bomb cyclone" (named for how rapidly it developed). Black is the highest humidity, red and yellow less so. Credit: NOAA

1. Fire Weather Tripled

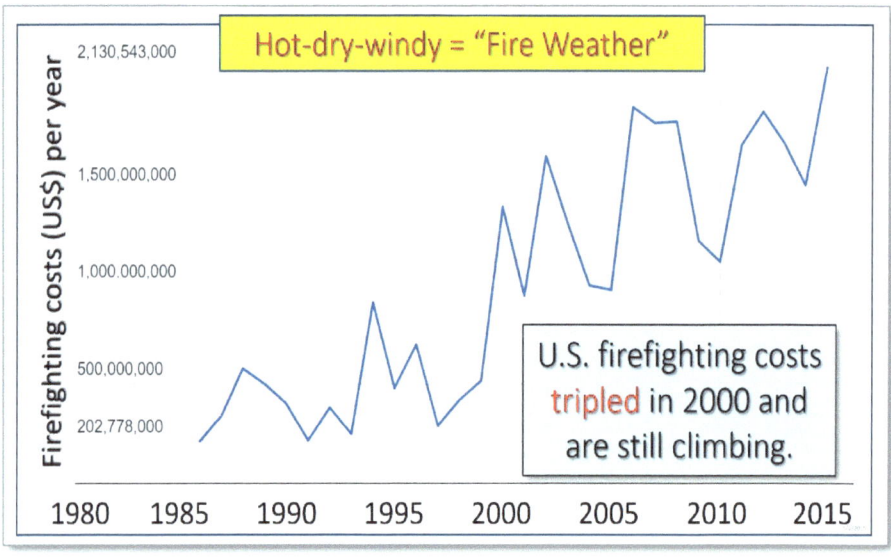

Figure 21. U.S. fire weather costs. Source: www.nifc.gov/fireInfo/fireInfo documents/SuppCosts.pdf

The term "fire weather" is used to describe a condition that is hot, dry, and windy—all three, together.

Big wildfire seasons are often included with extreme weather as they are an immediate consequence of hot-dry-windy conditions: all three, at once. The wildfire annual costs started creeping up after the global warming ramp began in 1977. In 2000, costs tripled and never returned.

- Firefighting costs tripled in 2000

"Fire weather" comes from adding **high winds** to the combination of two extreme weather types, **heatwave** and **drought**. Federal firefighting costs *tripled* in 2000 and have stayed up. They are now about **10X** times the 1990s baseline.

It's not that the *number* of fires has tripled; it's that more acreage burns, likely a matter of wind strength (the major concern voiced by experienced California fire fighters).

Wildfire burnt area in western U.S. forests, increase over 1973-1982	1983-1992	1993-2002	2003-2012
	640%	911%	1271%

Figure 23. This shows the increase in annual burnt **area** for U.S. wildfires in five large western forests, compared to the average of the decade 1973-1982 (which straddles the beginning of the global warming ramp in 1977). The following decade was up 640% and 2003-2012 was up 1271% over the decade when global warming began.

Are those numbers big enough for you? Here are some examples.

- The Dalles wildfire in 2018

Figure 24. The Dalles, Oregon, looking south from Washington State across the Columbia River.

Figure 25. The fire from above.

Figure 26. Looking south from that Dalles ridgeline several days later, showing six valleys that burned except where wheat fields were successfully protected by fire breaks.

- The 2018 'Camp' fire that destroyed Paradise

Figure 27. High heat from the 2018 Camp fire, which destroyed Paradise, California. The burnt vehicle's aluminum radiator and engine block have melted; only its steel parts remain. This shows that the temperature exceeded 400°C for some time. Credit: California Highway Patrol.

- San Francisco visibility in smoke

Figure 28. Emigrating smoke. Split screen LEFT: Looking southwest across the Bay Bridge from atop the Yerba Buena Island tunnel into San Francisco on November 16, 2018 during the Camp/Paradise fire, compared to a clear day a month earlier (RIGHT). The tall buildings are two miles (3.2 km) away; the distance to the second bridge tower, almost obscured by smoke, is 0.75 miles. The fire was centered 150 miles (240 km) to the northeast. Thanks to James R. Morrin, Jr.

2. Mega Heatwaves

Figure 29. Chicago humid heat wave of 1995. The excess deaths that week hit 739.

The long, hot summer of 1988 probably killed 10,000 people in the U.S. and Canada. That number covers about four months, as the extra heat started in May. But what made a "bad" heatwave "huge" in the 20th century was if it had peaks able to kill many hundreds of people in a short time. An example was Chicago's humid heatwave in 1995 that killed 739 people in only a week.

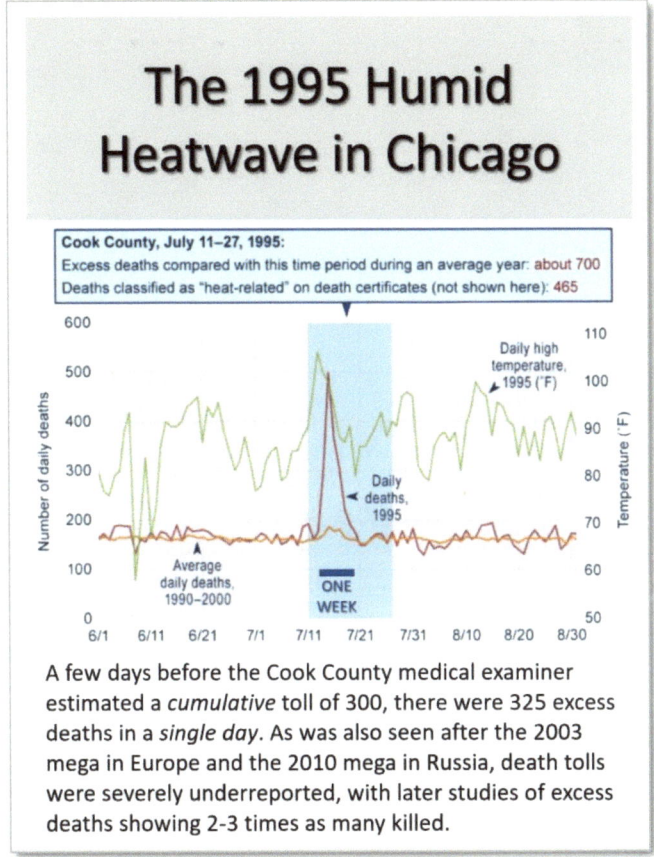

Figure 30. The spike in death certificates in Chicago.

This one is *not* a mega. It was only the big precursor. And a sign of things to come[7]: heat waves even more ferocious and lethal than the 1995 inferno will be annual events by 2080 unless current climate trends change.

In 2010 in Europe, there was a 5X surge in heatwave exposure, seen in the fig. 31 graph from the Lancet's annual review[8]. It too appears to be sustained.

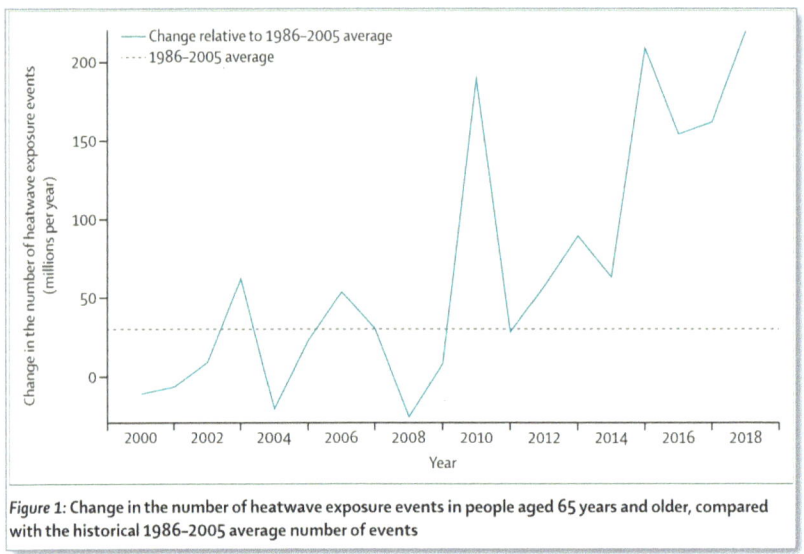

Figure 31. Heat wave exposures in the elderly.

In the 21st century, we got a big surprise in July 2003. There were 70,000 European deaths—those are "excess deaths" beyond the usual mid-summer numbers of death certificates[9]. This "mega" was described at the time as a freak of nature, one of those chance-in-a-million events.

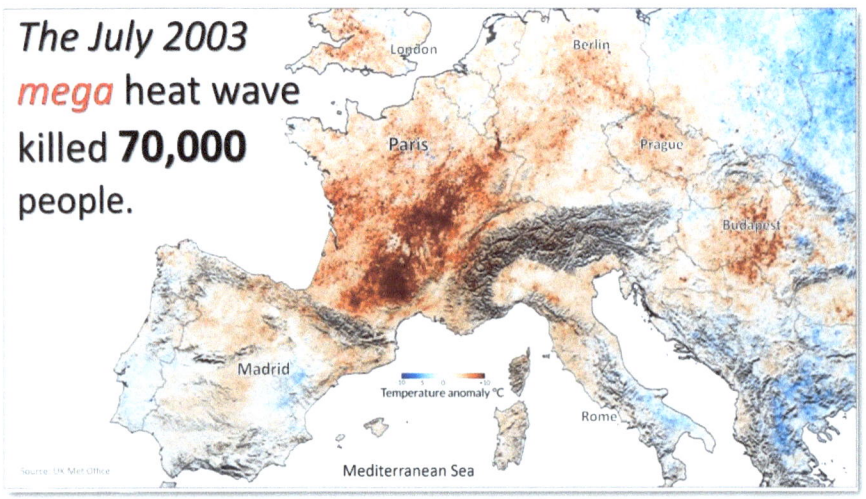

Figure 32. 2003 European mega's heat map.   Credit: UK Met Office

It was said that we were unlikely to see another in our lifetimes. Beware comforting words.

Nevertheless, a second mega occurred in 2010, hitting several time zones farther east in Russia[10]. It killed about 56,000 people in excess of usual numbers. So, a mega can quickly kill about **100X** more people than does a merely bad heatwave.

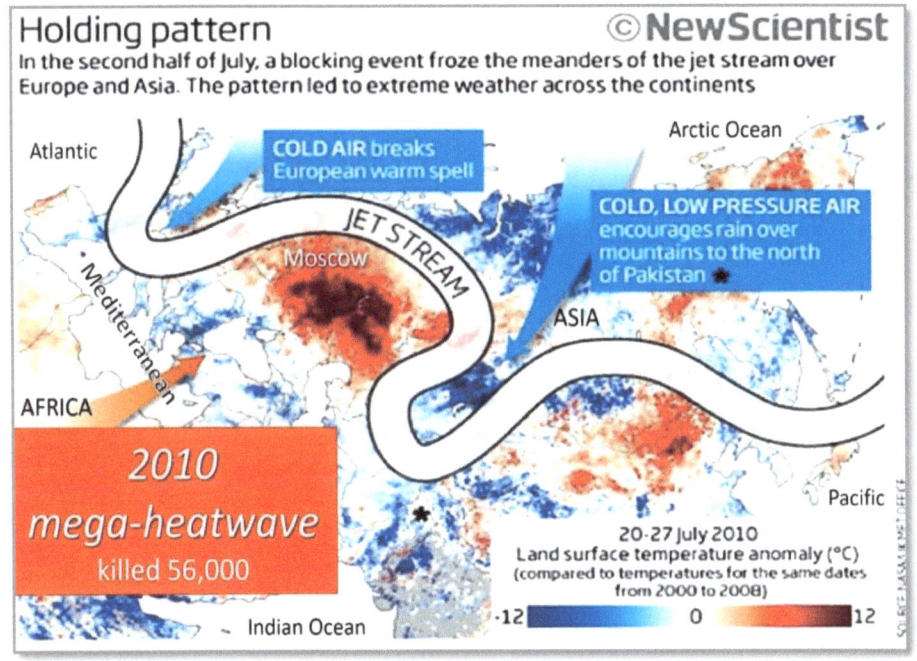

Figure 33. 2010 Russian mega. Credit: New Scientist

Extreme weather events are commonly analyzed in terms of *how intense* they are, or *how often* they occur, but often the most severe effects come from *how long* they last each time[11]. A long-duration event is, by definition, a condition that endures at least four consecutive days. On a more immediate time scale, two or three hot nights with no relief is thought to be an important factor in heat stroke deaths.

Prolonged weather conditions, where the weather just doesn't move on eastward, can lead to a variety of extreme weather events such as drought, heat waves, cold spells, and storminess[12]. For example, hot and dry conditions extended from April to September 2018 in western Europe, with only a few short interruptions of cooler and rainy.

There are effective actions we can still take to repel the extreme weather invasion, if only we get our act together in a hurry.

Like war, it is risky and uncertain. But properly focused actions can greatly improve our chances. The trip to Hell is not a sure thing.

2a. *An American Mega?* How Heat Kills.

Figure 34. The Kenya Marathon. That's Mt. Kenya in the background. Thanks to Martin Nyakabete.

Our meat-eating ancestors evolved, a few million years ago, to endure extreme exercise—and, furthermore, in the extreme heat of the African savanna. This was in aid of a simple hunting technique for grazing animals: running down prey in long chases on the shade-free savanna, with the beast having to stop now and then because of cramps. Once the hunter approached more closely, the beast would try to run again, limping. If the hunter first circled around before approaching, the beast could be driven back toward other hunters.

The hunters had better endurance than their prey, especially when there was a tag team of hunters that could switch around. They could apparently avoid immobilizing muscle cramps better. And we humans are usually able to keep our body temperature in the range where we function best—even under conditions such as the modern Kenya Marathon[13]. (Modern marathons really ought to be led by a mechanical zebra or wildebeest to provide the ancient inspiration.)

You'd think that we could live through our modern, increasingly frequent heat waves. So, what's the problem?

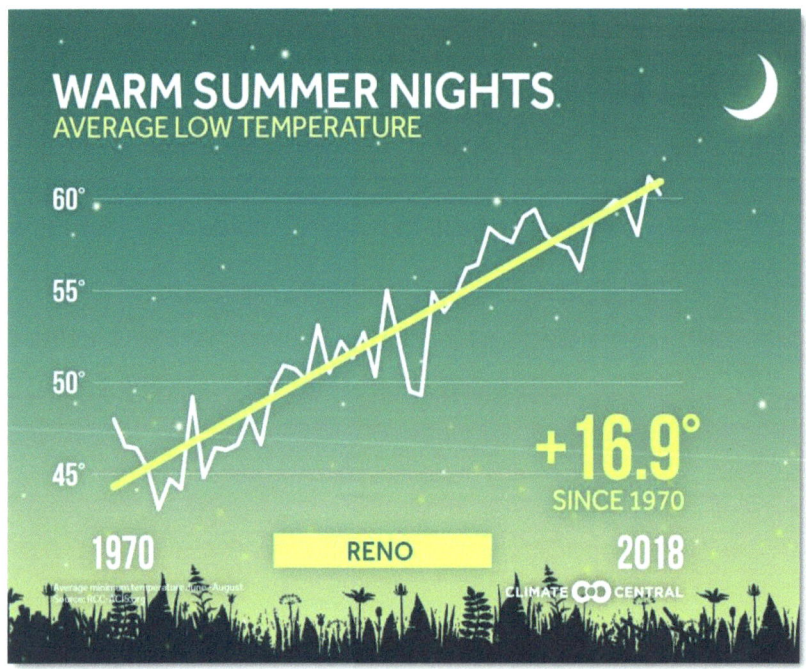

Figure 35. Reno's warmer nights. Some cities show little change.
Credit: Climate Central

. . .

Sweating and breathing both create evaporative cooling. Improving them is likely one reason that we became the barrel-chested naked ape, sometime in the last few million years of hominin evolution. But don't get me started; that's in the half-finished book manuscript that I set aside several years ago in order to write this book. The focus here is on how to survive the next twenty years, not how we managed the previous four million years.

So, how do heat waves kill? And where? Unlike the African savanna, the eastern half of the U.S. can be both hot *and* humid in the summer, as

I well remember from growing up in Kansas City and then going to Northwestern University near Chicago, hiking along the shore of Lake Michigan for three years as an undergraduate and then a year as a physics graduate student. Lake Michigan added to the humidity coming up from the Gulf of Mexico. Half of the physics department would play volleyball after lunch in the summer and I earned the nickname, "The anisotropic scatterer," as sweat kept getting into my eyes. When I got to MIT, along the Charles River looking over to Boston's Back Bay, it was just as humid.

When very humid, evaporation slows and can no longer counter the heat inputs. Sweat that drips off, or is wiped off, does not cool at all; it only dehydrates.

> **An aside:** Heat from a heatwave is nothing like running a fever. A fever is caused by your body's internal thermostat being turned up a few degrees, usually by an infection, commanding your metabolism to make more heat internally. This type of hyperthermia is, literally, under control. Your personal thermostat's set point merely changed to several degrees higher. Meds like aspirin turn the setting back down (don't take them if merely overheated).
>
> Fever is usually harmless; its evolutionary function, perhaps, is to make you feel bad enough to stay home (animals hunker down), and thus expose fewer people to the infection. If you bring down the fever with aspirin, stay home anyway. (Pathology with a social purpose!)

It is the heat source you cannot escape which can overwhelm the body's ways of cooling off. One of the better criteria for when *heat cramps* and *heat exhaustion* are advancing to the more dangerous *heat stroke* is when someone is no longer thinking normally. Confusion about time or place is the easiest heat stroke symptom to spot. (Step up and ask, "*Where* are you? *When* did you leave to come here?")

> **Take charge.** After assigning a bystander to phone emergency services for help, move the victim into the shade; if available nearby, onto the cool uncarpeted floor of an air-conditioned building (it will serve as a heat sink). Pour bottled water over the victim. Assign someone to update emergency services on the victim's new location and then go out to wave them in.

Dehydration is one common setup for heatstroke; it confuses both the heartbeat and the brain's higher functions. We are having a lot more long, hot nights in the 21st century. One is kept awake by the need to fan oneself and sponge to create more evaporative cooling. After two or three straight nights of no sleep, confusion sets in and heat-stroke fatalities climb.

Figure 36. Catching a breeze on a hot summer night, a century ago before there was enough electricity for air conditioning. Credit: U.S. Patent Office.

It's not just a matter of falling down the stairs. Failure to cajole the more vulnerable youngsters and seniors into drinking water every few hours can set them up for dehydration and its fatal effects on heart rhythms.

The episodes of extremely hot nights have become more widespread since the global warming ramp began in 1977.

But even when dehydration is avoided, extreme heat can kill in other ways. Tissues overheated for too long become inflamed, dropping blood pressure and slowing oxygen delivery. Most patients hospitalized for heatstroke show signs of systemic inflammatory response syndrome. Much like septic shock, it is a vicious cycle that can eventually cause various organs to stop working, killing the victim.

Neighbors should organize in advance, so that the able regularly check up on the less able, leaving behind cool six-packs of bottled water for bedside consumption. When a series of hot nights is forecast, opening generator-powered cooling centers is not enough. Elected officials should promptly declare a mutual-aid emergency, one that brings in many outside trained medics and fire-fighters to assist. Locksmiths, too—though the fire-fighters have their own ways of entering to check up on people who don't answer their phone or doorbell.

Despite a few preparations in a few places, heatwave remains the leading cause of weather-related death in the United States—not the windstorms and floods featured in the media.

In the meantime, do remind your elected officials to prepare for an American mega and do something to back out of the danger zone for extreme weather.

Figure 37. Looking on the sunny side.
Thanks to *du Monde*.

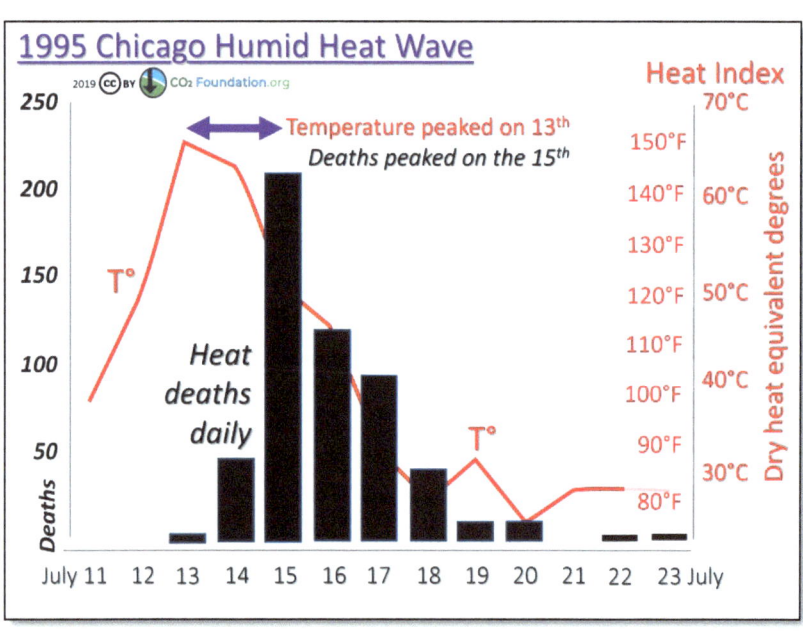

Figure 38. Note the two-day lag between stress and death.

3. "Those hurricanes that don't leave"

Hurricanes are also known as typhoons in the western Pacific and Indian Oceans, and as tropical cyclones or 'TCs' in science and in Europe. Upon making landfall, they keep going at an average 11 mph (17 km/h) ground speed and so limit their visit to a coastal city to less than five hours.

While one can argue about whether global warming has increased the numbers or sizes of tornados and hurricanes, it is now clear that *ordinary* hurricanes, *when stalled*, are becoming a big threat. There are other ways in which to be big, such as staying 24 times longer than when merely passing through.

Here are a few examples of when hurricanes stalled, all very expensive (your tax dollars and insurance premiums at work, even if you live inland).

• The path of the 1991 "Perfect Storm"

The 1991 Perfect Storm was a nor'easter, which then evolved into a small hurricane. It started, oddly enough, off the coast of Atlantic Canada on October 29, right on the great circle route from East Coast cities to Europe used by both shipping and air travel. Forced southward by the jet stream blocking its way to the usual northeast, it reached its peak intensity as a large and powerful hurricane.

The east coast of the United States was hit with high waves and coastal flooding before the storm turned to the southwest and weakened. Moving over the warmer waters of the Gulf Stream, it became

Figure 39. The 1991 Perfect Storm wandered offshore. The points show the location of the storm at six-hour intervals. Note how it slowed down (closer spacing) during the 270°-left-turn loop.

a subtropical cyclone. It executed a slow 270° turn off the Mid-Atlantic states and headed northeast, evolving into a full-fledged hurricane, with peak sustained winds of 75 miles per hour (120 km/h).

Not having located a more authoritative source, I have marked the beginning of stalled hurricanes at 2012, with Sandy, and am treating this event 21 years earlier as a precursor.

- Hurricane Sandy in 2012 (a forced left turn)

Figure 40. Coastal erosion at Mantoloking, New Jersey. Hurricane Sandy high waves cut through this narrow peninsula. Credit: Greg Thompson, USFWS.

Hurricane Sandy[14] was, in 2012, the largest Atlantic hurricane on record, as well as the second-costliest Atlantic hurricane in history, surpassed only by New Orleans-bound Hurricane Katrina in 2005. At least 191 people were killed along the path of Sandy, in seven countries.

A jet stream loop blocked the usual path to the northeast taken by offshore hurricanes, and so it turned west, coming ashore at Atlantic City. Sandy was so wide that New York City was hit hard by the sea water pushed up the Hudson River by the winds.

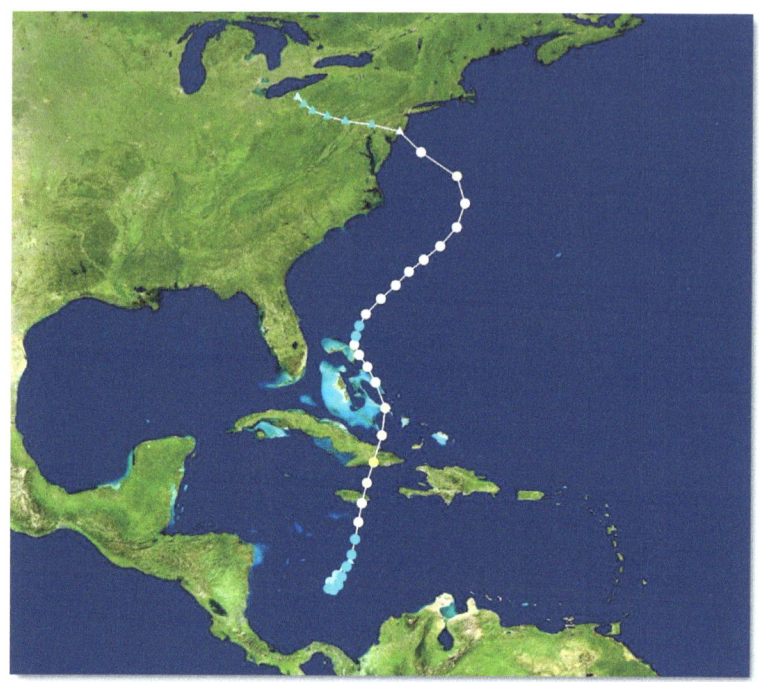

Figure 41. Track of 2012 Hurricane Sandy. Six-hour spacing.

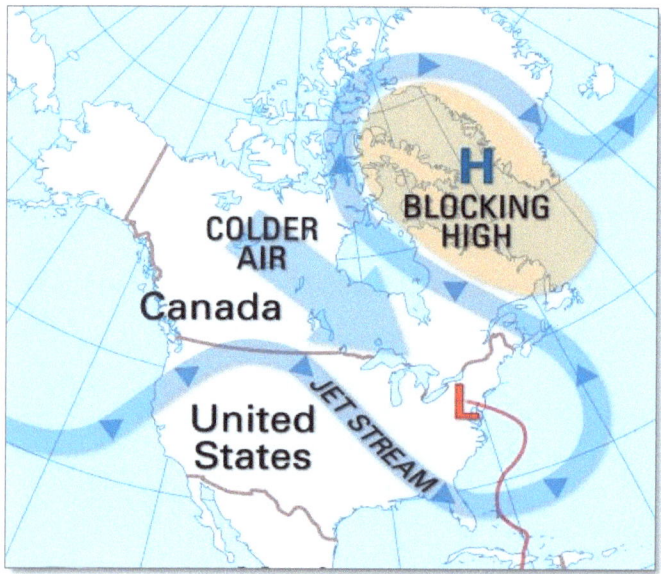

Figure 42. The polar jet stream configuration that forced 2012 Hurricane Sandy (red) into a left turn over land. Credit: Greene et al.

- Hurricane Harvey (visited Houston for a week in 2017)

Hurricane Harvey in the summer of 2017 became our most costly example of stalled severe weather. Not only did the Houston area get a year's worth of rain in only four days, but the storm surge pushed saltwater inland, preventing the rain from running off.

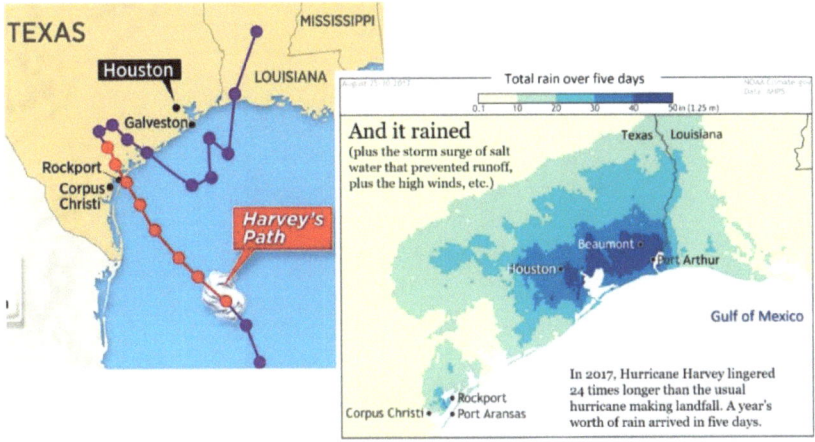

Figure 43. 2017 Harvey track and its rain map

Figure 44. 2017 Port Arthur TX flooding

● Hurricane Dorian in 2019

Had Dorian's path shifted a hundred miles west, this supersized hurricane would have set all-time records for damage done. As it was, we merely got "a warning shot over the bow."

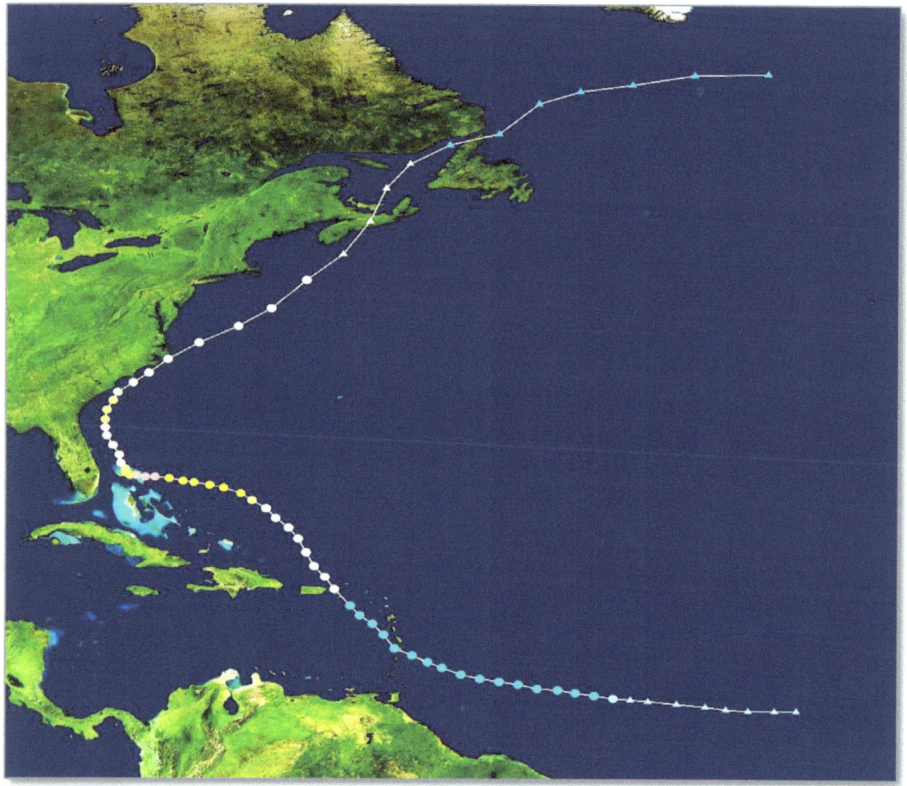

Figure 45. Some hurricanes just slow down, as did the supersized Hurricane Dorian in 2019 after turning north over the Bahamas (six hours between dots, so advance is slow when dots touch, nearing Florida).

4. Severe inland windstorms are up 8X

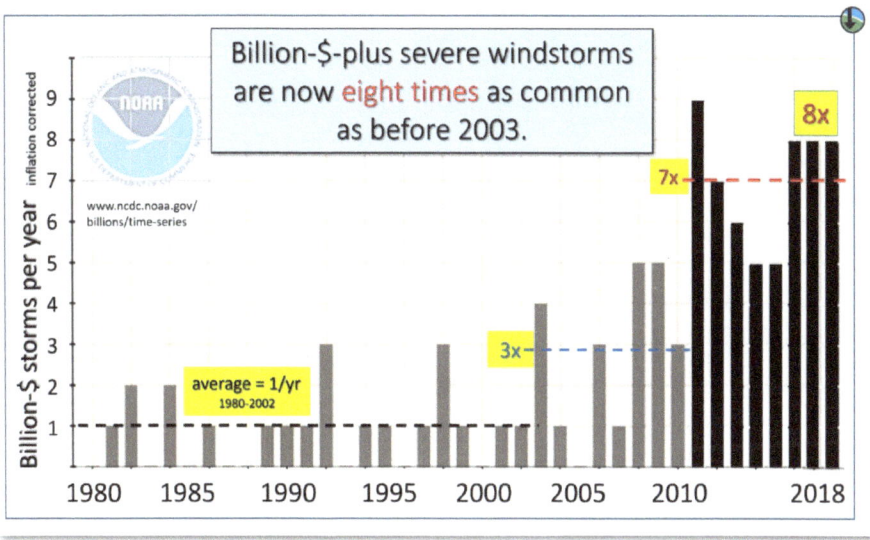

Figure 46. Surge in severe windstorms (NOAA counts tornados, derechos, hailstorms, but not hurricanes). This plots only the events that did more than US$1b in damage (some, much more).

In the good old days (1980-2002 in the case above), the U.S. averaged only one windstorm per year that managed to do more than a billion dollars in damage, correcting the older ones for inflation.

After 2003, the annual number of billion-dollar-plus storms *tripled*.

After 2010, we have averaged seven each year. Then we had three successive years of *eight* billion-dollar-plus windstorm events.

Tornado, derecho, and hailstorm are included in the count—but *not hurricanes*. So, these are *inland* billion-dollar-severe windstorms [15] whose numbers increased **8X** beyond their pre-2003 numbers.

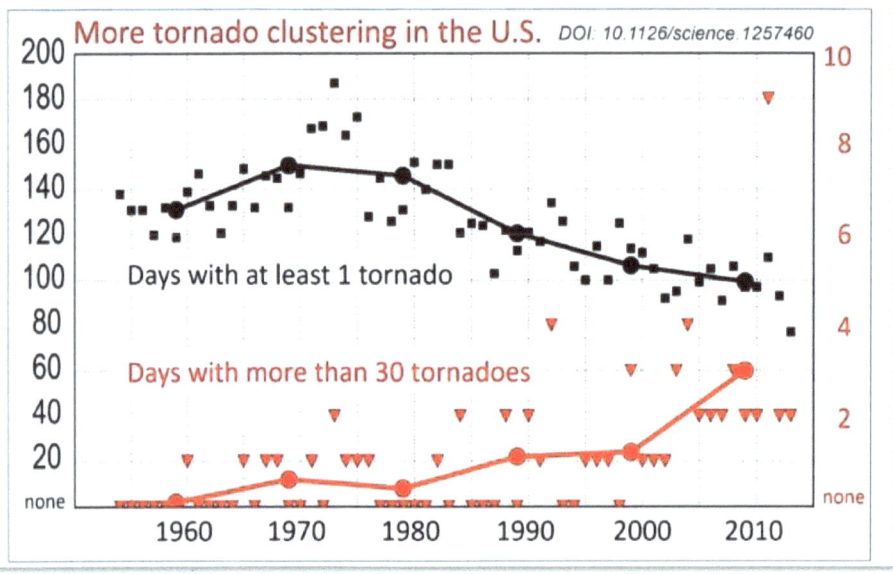

Figure 47. Tornado clustering

Speaking of tornados, they began to cluster after 2000: there are now fewer days in the year with a tornado somewhere in the U.S.—but many

Figure 48. Tornado. Credit: NOAA.

more days when 30 or more tornados are reported. Derechos also cluster.

Figure 49. Railroad bridge in New Mexico. Blown away, train tracks and all, indicating that the wind pushed *two-high cargo containers* sideways until the track's framework (that boxy grid in the foreground) broke loose from supports. This serves as an example of why rebuilding infrastructure is essential. Credit: New Mexico State Police.

Figure 51. "Tornadoes have been popping up every day in the U.S. as if coming off an assembly line. They're part of an explosion of extreme weather events, including record flooding, record cold and record heat." *Washington Post*, May 29, 2019; Dayton, Ohio. (Credit: John Minchillo/AP).

Figure 50. Waterspouts in Lake Michigan, 2013. Credit: Kenosha Police Department.

4a. Familiar with the Derecho?

"Is that a rift in space-time, Daddy, or just a cloud?"

Figure 52. This Sydney squall line in Spring, 2014, had winds of more than 160 km/h (100 mph, likely from downbursts) that blew down trees and power lines, leaving 30,000 homes without electricity. Thanks to Nick Moir (that's his daughter).

To summarize, a derecho is fast-moving, gives little warning, and the downbursts are so loud that you won't hear the nearby trees falling. A derecho's path can be both wide and long.

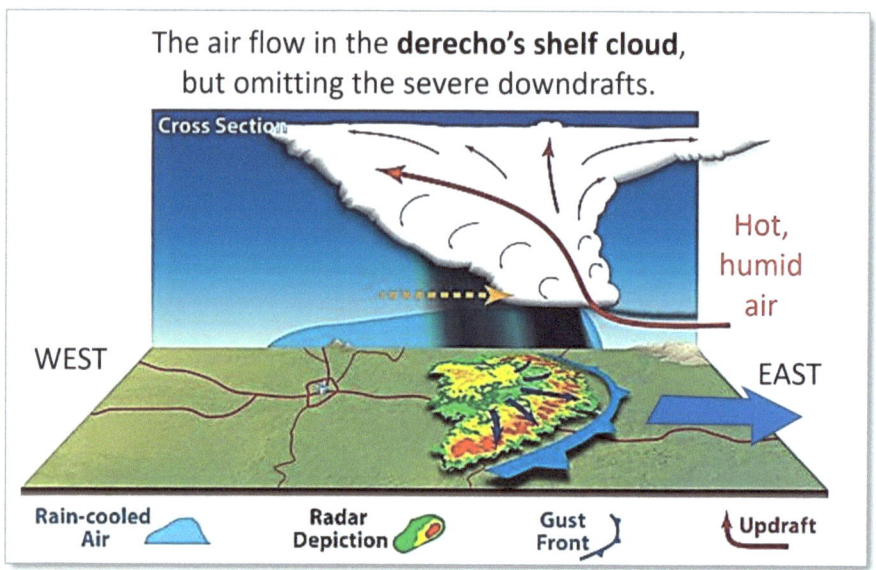

Figure 53. Cross section of a derecho.

Usually severe winds spiral around, as in the tornado and hurricane. The derecho ("deh-REY-cho" for us Anglophones) is a rarely seen type of severe windstorm, named 'straight' in Spanish for its severe winds out of the east.

The U.S., to the east of the Rocky Mountains, has developed this form of climate disease earlier than has the rest of the world, so it is worth a look at the derecho progression in the U.S. to see what may be ahead for other places.

Given how infrequent derechos have been, we are lucky to have a first-hand account from a trained meteorologist. Sarah Jamison, a National Weather Service meteorologist, was enjoying the 4th of July holiday weekend in 1999 at a campground in western Maine (in fig. 54, look for the red **X** at right) when North Dakota weather reached out and touched her after a speedy 15-hour trip through Canada. Here's the debrief[16].

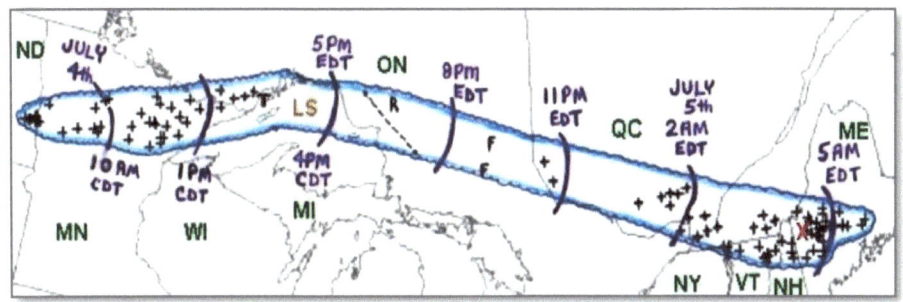

Figure 54. Track of the 4th of July derecho in 1999.   Note that red **X** above.

July 4th was a very hot and humid day, with an afternoon high near 90°F (32°C). As the night progressed, Sarah mentioned that "the air was very stagnant with no wind, making it very uncomfortable in the hot tent."

Awake, she noticed lightning in the distance. Quickly, the gust front hit and the wind rose alarmingly. Once everyone had run to the SUV in the parking lot, they could see trees falling all around with every lightning flash. About two dozen trees were blown down, several of which were larger than two feet (60 cm) in diameter.

After about five minutes, the winds began to weaken; within a half-hour, the storm had passed. One of the large trees and two smaller ones had fallen on Sarah's tent, completely crushing it. Fortunately, no one in the group was hurt.

The roar of the storm's winds was so loud that none could recall having heard the trees snapping or crashing to the ground. As is often the case with derechos, the damage that hit Sarah's campground was associated with a narrow band of intense, very damaging downbursts. The derecho continued to cause damage across central and southern Maine. Finally, after traveling over 1,300 miles (2,100 km), the derecho ended just before reaching the Atlantic coast. It had begun 21 hours earlier in North Dakota.

To summarize, a derecho is *fast-moving*, *gives little warning*, and the *downbursts* are so loud that you won't hear the nearby trees falling. A derecho's path can be both wide and long. I hope to never see one.

And the derecho is no longer rare—sometimes, another arrives the very next day. And maybe the following day as well, as happened to

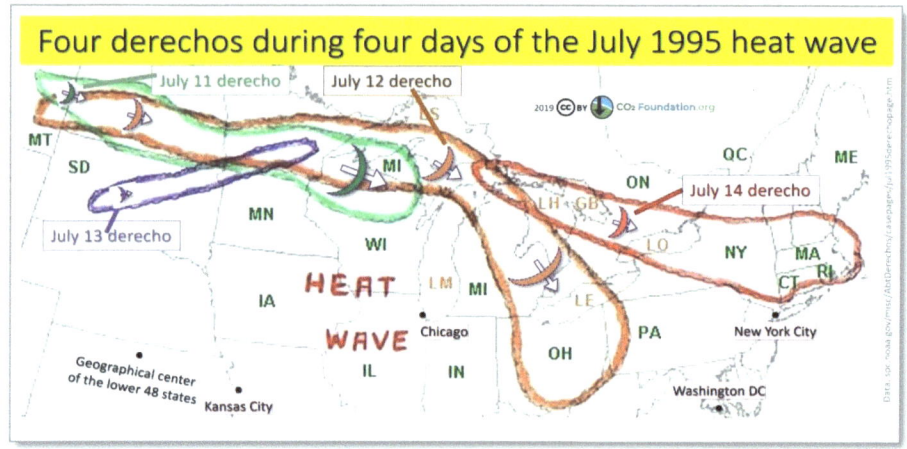

Figure 55. July 1995 derecho paths, back in the days when color graphics were drawn with colored crayons.

Minnesota (MN) in 1995. Worldwide, a crude count of derechos suggests a ten-fold increase in annual numbers after 2006.

Derechos are big monsters. To qualify as a derecho, a windstorm needs to produce a continuous swath of wind damage that is (at least in the U.S. definition) longer than 400 km (250 miles). This hypothetical derecho in England shows the scale.

- ## The "Inland Hurricane"

In the U.S., the derecho is sometimes called an "inland hurricane" because of its high winds, but the derecho covers ground five times faster than a hurricane and twice as fast as a tornado—more like a fast freight train at 60 mph (100 km/h). They travel eastbound, though sometimes one is forced to detour to the northeast or southeast.

It's not just *faster* ground speed and *longer* reach: a derecho also cuts a far *wider* swath of damage—say, 250 miles (400 km) wide (that's London to Scotland) rather than the 50 miles (80 km) for a hurricane. The biggest tornado is less than a mile wide (1.6 km).

Hot and humid weather is the usual setup for a derecho, and we now have more hot and humid weather to concern us. Derecho numbers[17] (no

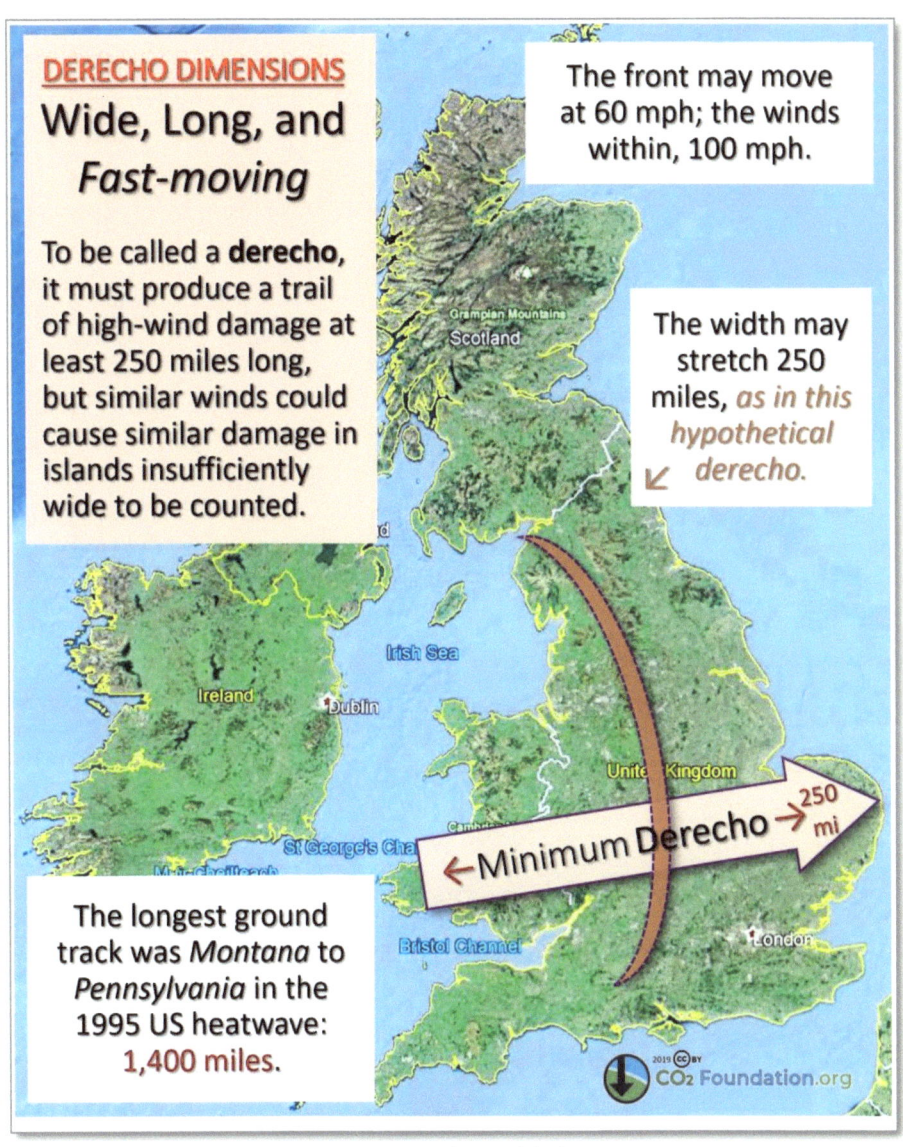

Figure 56. Derecho width diagram.

billion-dollar threshold this time) are up **10X** since 1990, the big step occurring in 2006.

Derechos have some tendency to cluster, thanks to requiring a heatwave to get them started.

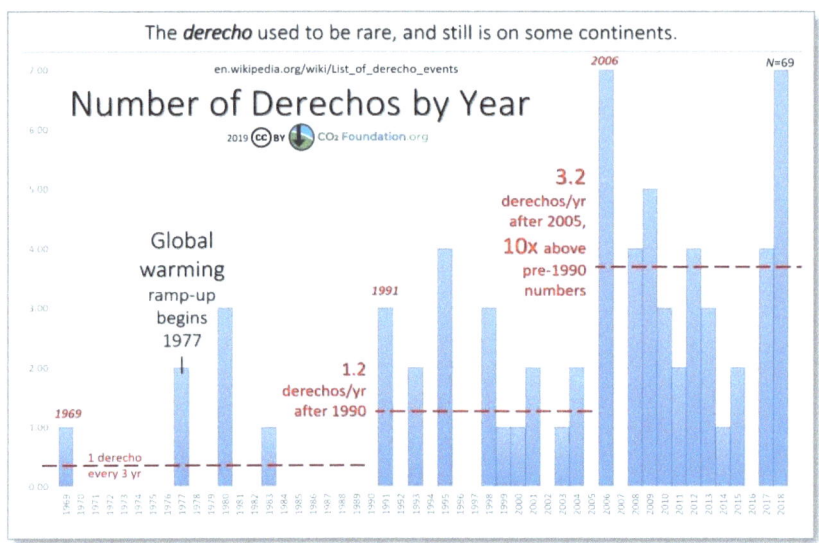

Figure 57. Plot of worldwide derechos per year.

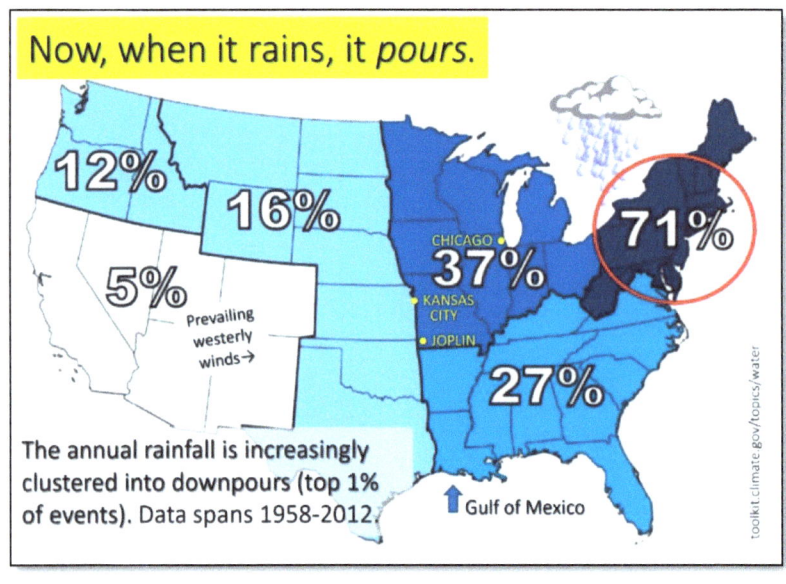

Figure 58. Downpour events have increased, but mostly in the eastern half of the continental U.S. That's also where most of the severe weather concentrates.

5. Severe inland floods up 4X after 2009

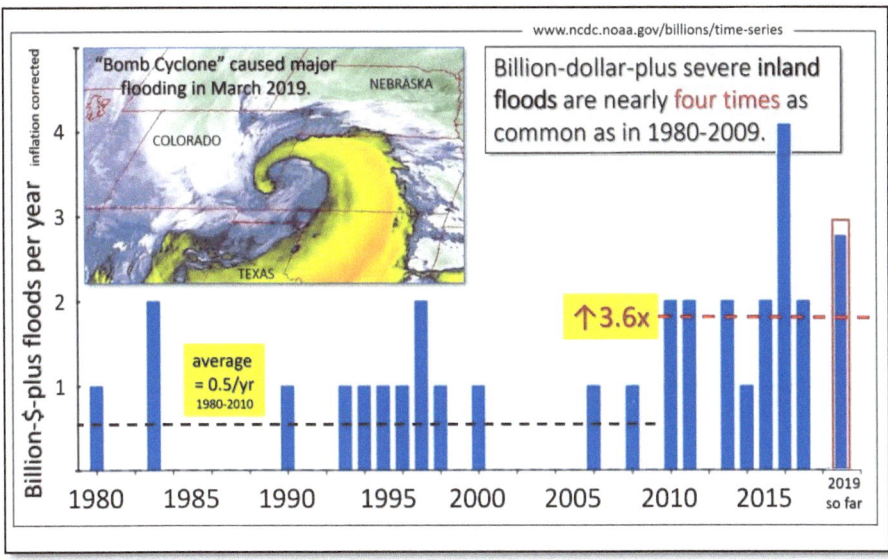

Figure 59. Since 2010, severe inland floods are up about 4X; this does not count coastal flooding from hurricanes. The **inset** shows the central portion of that March 2019 'bomb' cyclone. It caused one of the three billion-dollar floods of 2019.

The annual number of big inland floods, each topping a billion dollars in damage, took a big step up in 2010, to 3.6X the prior average (again, hurricanes are not counted).

But our trouble with prolonged heavy rain started earlier when it began clustering into heavy downpour days, much as tornados have been clustering. Annual numbers fail to capture the action, as in this example of many more downpour days.

Five times more dengue virus circulating

It's not weather per se, but in the U.S. there has been a 4X increase in dengue virus vectors from wind-distributed insects. That is likely a consequence of the surge in severe windstorms, not just the temperature creep.

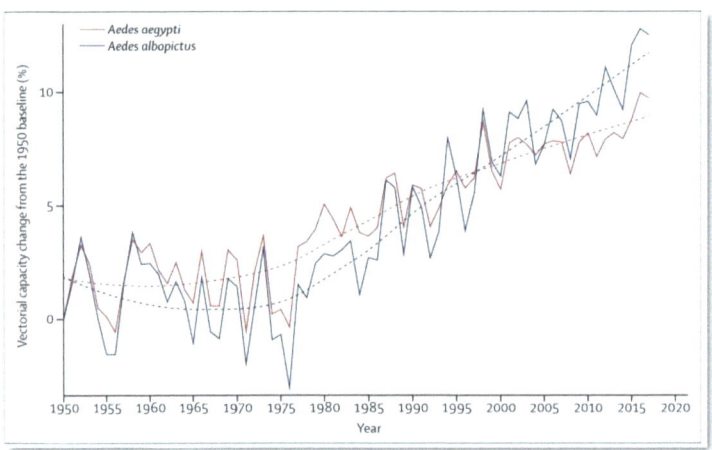

Figure 60. More circulating dengue. Credit: Centers for Disease Control

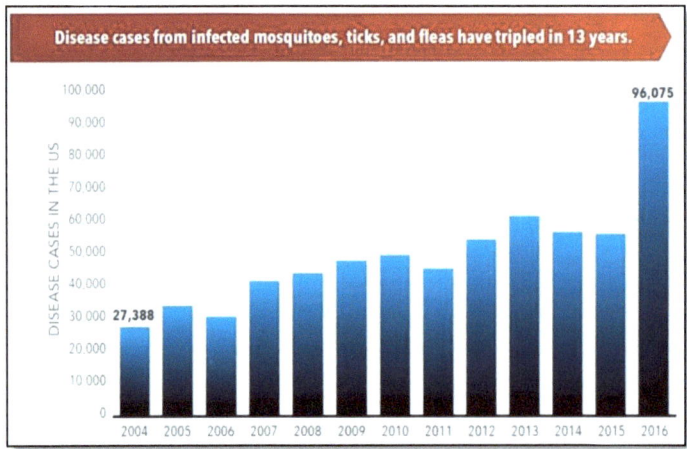

Figure 61. Windborne disease is up 3X in 13 years, likely aggravated by more billion-dollar windstorms.

Worse, but not tripled: long-lasting droughts

Figure 62. Texas, 2016. Notice the green plants in the foreground. Dry is relative and so is drought. The maize ('c o r n' in the U.S.) crop has failed; it is a risky crop because of its need for a lot of water. Had the farmer planted beans, a good dry crop, the food supply likely would not have been affected. Adaptation involves just such choices, sometimes foregoing maximizing this year's profit; one gets a more assured return from a mix of crops.   Thanks to the USDA's Bob Nichols.

Some pre-1977 droughts in the U.S. lasted 20-30 years and were also widespread, creating major hits on agricultural productivity. That happened four times within a hundred-year span, even back before global warming[18]. Our global overheating only worsens that historical threat.

In other parts of the world, the proximate cause can be identified. In the Indian subcontinent, failure of the monsoon rains is feared. In an El Niño, the trade winds become disorganized; they may fail to bring ocean moisture ashore for several years. This is what, in medicine, we would call the 'natural history of the disease,' the baseline fluctuations we must use to judge if a treatment is effective or if it merely overlapped with a down fluctuation in symptoms.

Even without the new overheating and extreme weather, there's a lot of drought threat for the U.S. in the 21st century. A common cause has not been identified for multi-decadal drought, though changing winds off the Pacific Ocean are suspected to play a role.

However, **seasonal drought** (low rainfall in the Spring growing season) seems to be a setup for the mega heatwaves. And we do have a candidate cause for those droughts: jet-stream meanders associated with big high-pressure cells that are slow to move on. Too many cloudless days spells low rainfall.

Figure 63. Drought in Ethiopia in 2019.

Worse, but not yet triple: "Crazy ice storms"

Figure 64. Unseasonable freeze. Icicles and frozen ripe nectarines from freezing rain in Lakeland, Florida, in 2010.

Figure 65. An April frost.

Though freezes may not damage fruit trees when they are dormant, with their 'pipes' drained for the winter, freezes during the growing season may kill the tree via cracking the pipes, thus preventing leaf evaporation from pulling up more water and nutrients during the growing season.

New fruit trees can take ten years after transplanting to become productive, so an out-of-season freeze may eliminate the next ten years of crop. Fruit farming is becoming a much riskier business, thanks to jet stream meanders.

The 21st-century version of extreme weather is not just the far end of the 20th-century statistics, creeping higher.

We are seeing something new—the 21st century surge was sudden and sustained—and the threats are growing rapidly.

DIAGNOSIS
(identifying causes)

Figure 66. The *Kink*.

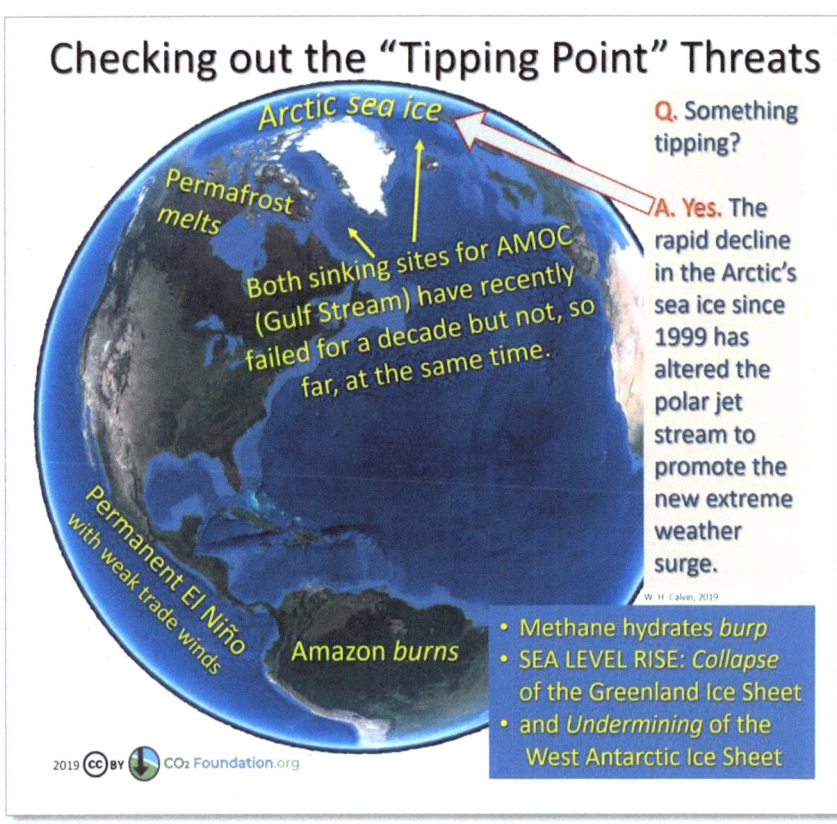

Figure 67. Identified "tipping points." The sea ice loss has accelerated since 1999, warming the Arctic and rearranging the winds, especially the polar jet stream. Once past a tipping point, things get worse automatically and the situation is difficult to reverse within a thousand years.

Kinks in the jet stream

A broad U-shaped loop of the polar jet stream can be squeezed into a hairpin turn by the failure of the leading edge to advance eastward—while the trailing edge continues moving.

Such hairpin turns containing cold Arctic air have been known to sit atop the entire U.S. East Coast, freezing growing fruit in Florida and allowing propagandists to mislead people by claiming there is no global warming (and thus no reason to reduce fossil fuel use).

I doubt that out-of-season freezes have recently tripled in numbers, but I am appending them to the Big Five because they share a common cause and because of their outsized effect on public skepticism about the global warming explanation for our extreme weather problems. "Warming causes cold snaps?" they say. "That defies common sense." But scientists love such a paradox and start trying out explanations for size and fit.

A big stumbling block, at least in the popular mind, has been the seeming paradox of freezing cold caused by warming[19]. Let me start by noting that the freeze is temporary and *regional*, while the warming is a much smaller, long-lasting *global* change. Beware of people unintentionally comparing apples and oranges.

Perhaps we should take a look at the list of known tipping points for climate. Seven of them are illustrated in fig. 67.

When the Jet Stream Wanders

Climate scientists started looking for new players, especially from the list of potential climate tipping points they had been compiling—catastrophes such as burning off the Amazon rain forest. Or a shutdown in what we now call the Atlantic Meridional Overturning Current (*AMOC*, to avoid people calling it the Gulf Stream[20], which is merely one segment of the AMOC's circuit).

The two major ocean sinking zones, near SW and NE Greenland, have each shut down in recent decades, and for years at a time[21]; had their shutdowns overlapped in time, the entire planet would have been severely affected. That's what I wrote about in 1998, in the first major magazine article on "The Great Climate Flip-flop" (as *The Atlantic* editors titled it).

Both the rain forest and the AMOC strength are declining—bad news in the long run as they both have thresholds for total collapse—but little connection has been found to our present extreme weather surges.

However, further down that tipping-point threat list was the rapid warming of the Arctic, pronounced after 1993. The rapid loss of Arctic sea ice after 1999 has caused the polar jet stream to swing south more frequently[22]. It then began snaking around the mid-latitudes in a more dramatic fashion, just as a river will develop meanders when it slows down in a flatter landscape. The meanders turn out to have a great deal of explanatory power, consistent with promoting many types of extreme weather.

The path itself is called a standing wave. The path slides eastward like a snake glides forward. In the 20th century, the jet stream usually blew out of the west at 100-200 mph along a mildly sinuous path, up at about 33,000 feet (10 km). In flying from Seattle to New York City, one could arrive an hour early if the pilots picked up a tailwind from the jet.

Figure 68. Napa River meanders in 2006. Flying over the Amazon rainforest, one often sees rivers meandering. Those lakes off to the sides (*abandoned meanders*) show where the Napa River in Peru used to run a few decades earlier, back before a flood cut a new channel that served to shortcut the longer path. Silt piles up on the new curve to isolate the old loop, which becomes a lake. In the U.S., the Mississippi River has impressive meanders near where the Ohio River joins.

The trick was avoiding the headwind on the return trip, often requiring a longer-but-quicker flight path up over Canada.

The jet's path has always wandered somewhat, with perhaps four topside turnarounds along the sinuous round-the-world path through the upper mid-latitudes, covering perhaps 15,000 miles in circling the earth. The standing wave's amplitude (and thus the total path length) often increases during the winter and dampens down in the summer.

The turnarounds of this standing wave slowly drift eastward, much like that gliding snake; the southern turnaround of the path might drift across North America within a week or so, moving the Highs and Lows with it so that any given locale gets an alternation of sunny skies and rainclouds. But sometimes they stall and fail to continue east. That lengthens the loops into hairpin turns and may create additional loops. Like kinks in a garden hose, they cause trouble.

> During the extreme events I noted, the jet stream acted strangely. The bends went exceptionally far north and south, and they stalled—they did not progress eastward. The larger these bends, the more punishing the weather gets near the northern peak and southern trough. And when they stall—as they did over the U.S. in the summer of 2018—those regions can receive heavy rain day after day or get baked by the sun day after day. Record floods, droughts, heat waves and wildfires occur.
>
> My collaborators and I have recently shown that these highly curved, stalled wave patterns have become more common because of global warming, boosting extreme weather.
>
> —climate scientist **Michael Mann**, in *Scientific American*, March 2019

Figure 69. Jet stream kinks

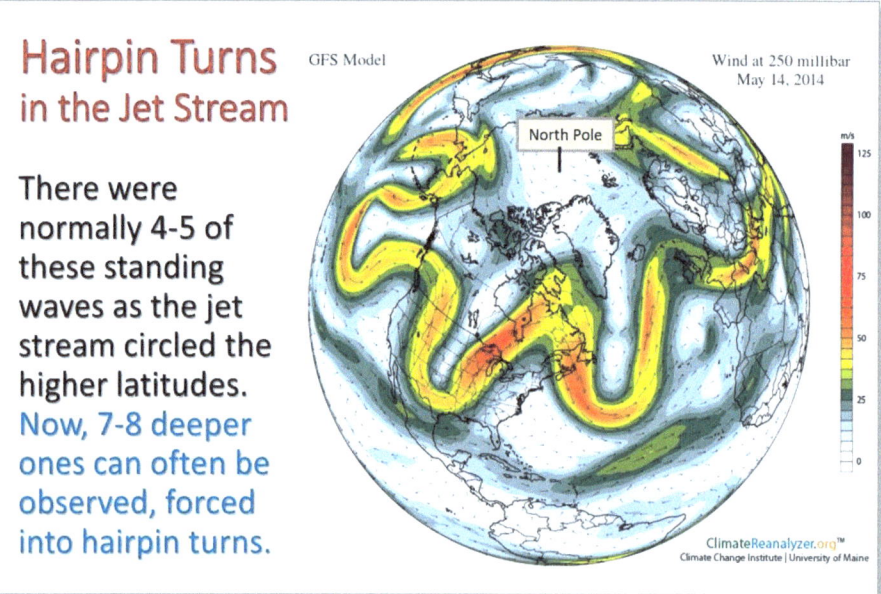

Figure 70. Hairpin turns along the standing wave

Detached Highs & Lows Spinning Down

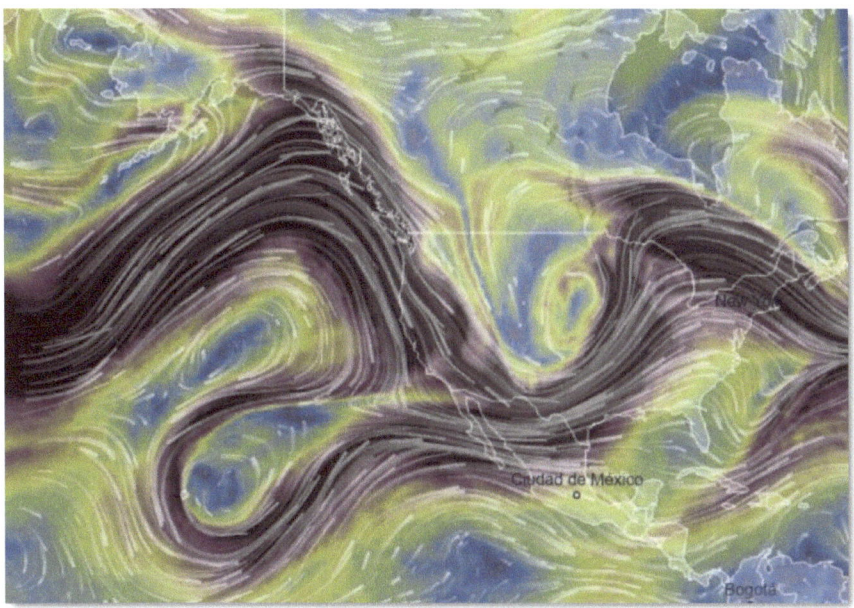

Figure 71. 2019 model for the polar jet stream going south along the U.S. west coast, then splitting off a loop heading for Hawaii. Such loops may close off and detach, taking a week to spin down.

In addition, the end of a loop may become detached when the loop begins shortening as the jet stream finds shortcuts. And so detached highs and detached lows are closed spirals that take a while to spin down. They look like those abandoned meanders from former paths of a river, sealed off after a spring flood.

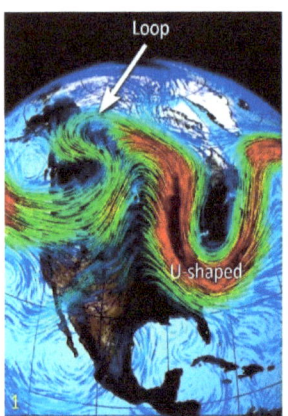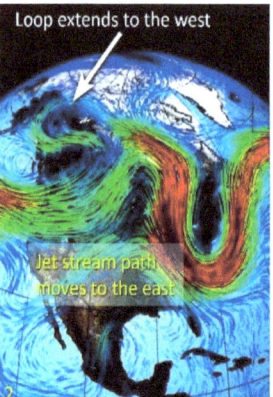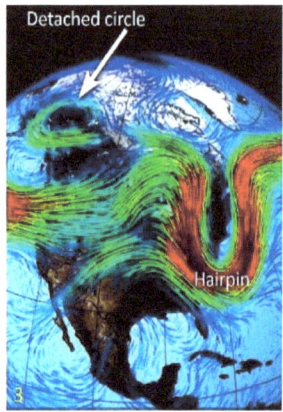

Figure 72. Hairpin plus detaching loop over Alaska. A time-lapse series of the polar jet stream progression during the June 1988 heat wave (watch at *svs.gsfc.nasa.gov/3864*)

1. The jet comes off the Pacific Ocean into the Pacific Northwest, turns north into Alaska, and then makes a U-shaped dip into the southern states before heading north up the Atlantic Coast. But notice that loop (white arrows) into Alaska.

2. The standing waves slowly move eastward but notice that the right side of the U is beginning to stall, causing the broad U to be squeezed.

3. The U has now become a hairpin. Note also that the Alaskan loop has pinched off, forming a "detached low," another contribution to 'crazy weather.'

The Strange Year of 1988

The drought of 1988 ranked as the worst U.S. drought since the 1930s Dust Bowl, with damages (inflation-adjusted) of about $100 billion. During that long, strange summer, heatwaves killed about 10,000 people in the United States and Canada; drought affected 23 of the lower 48 states.

In the three screenshots of the polar jet stream in Figure 72. Hairpin plus detaching loop over Alaska. A time-lapse series of the polar jet stream progression during the June 1988 heat wave (watch at svs.gsfc.nasa.gov/3864) covering June 10 to July 8 of the 1988 heat wave in the Midwest. That was only 11 years into the global warming ramp and 4 years after the land surface began warming 150% faster than the ocean surface. So, 1988 may serve to mark the beginning of our new extreme weather—or, at least, a precursor.

Figure 73. Dr. James E. Hansen, Congressional testimony in 1988

In the midst of it, NASA climate scientist James E. Hansen told a U.S. Senate committee he was 99 percent certain that the year's record temperatures were not the result of natural variation. It was the first time a lead scientist drew a connection between human activities, the growing concentration of atmospheric pollutants, and a warming climate. The future that was forecast in the 1960s was already happening in 1988.

"It's time to stop waffling so much and say that the evidence is pretty strong that the greenhouse effect is here," Hansen told reporters.

Thirty years later, not much has been accomplished by the U.S. Congress, showing the power of heavily promoted climate denial.

...

How much CO_2 do we need[23] to quickly remove? For that, we now have some track record to give us a rough estimate.

- In the mid-1970s when the warming ramp began, the CO_2 concentration had risen from the preindustrial 280 ppm to 330 ppm, an 18 percent excess.
- In 1988, the CO_2 concentration in the air overhead had reached 350 ppm, a 25 percent excess. Holding emissions down, so we don't exceed that, became the target at *350.org*.
- At 420 ppm in 2020 or 2021, we have an excess of **50 percent.** We are adding to it at a rate about five times faster than in the 1960s.

That suggests to me that we will have to *remove* most of the excess CO_2 to have a chance of restoring the former climate. And, judging from our extreme weather surges, we now have very little time in which to do it.

Figure 74.

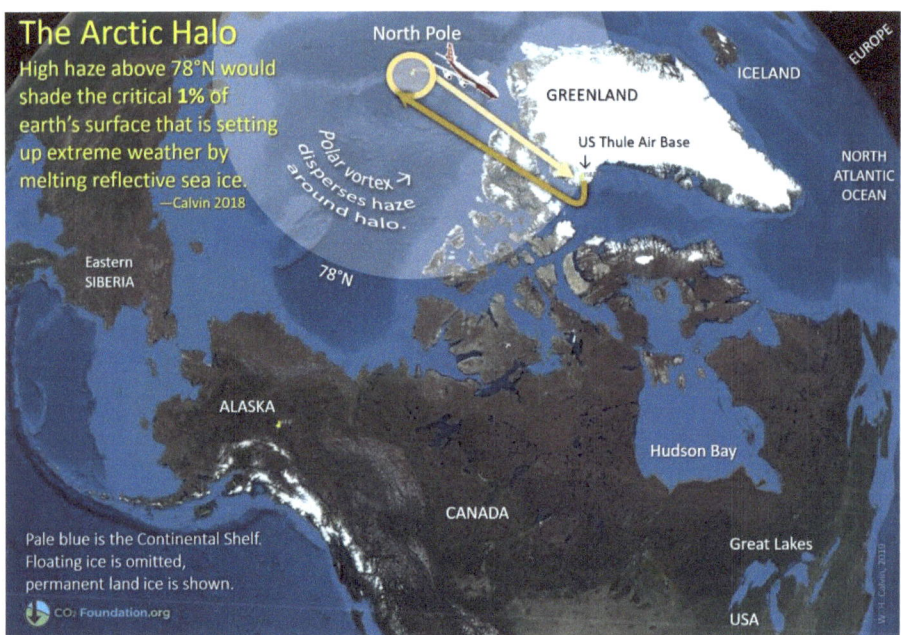

Figure 75. The High Arctic may be only 1% of the earth's surface, but its floating ice plays a critical role in climate elsewhere.

What's Behind Extreme Weather?

Figure 76. Five in the chain of causation for extreme weather ("X-Wx").

Severe windstorms, prolonged downpours, mega heatwaves, much more fire weather, and stalled hurricanes—all of them surged between 2000 and 2012. None have retreated.

It's time to start saying that we had an *abrupt climate shift* in that dozen-year period.

I realize that earlier abrupt climate changes were characterized by jumps in surface temperature and in precipitation within a decade or two—but that is because it was the only data we had to judge a climate shift, on the time scale we choose to call "abrupt" rather than "rapid" or "fast." Little else[24] was available from tree rings and ice cores on a decadal time scale.

Severe windstorms, prolonged downpours, mega heatwaves, much more fire weather, and stalled hurricanes—here we have five types of annual data that are conventionally included under the rubric of climate change. Each has quadrupled within a decade—and stayed high.

It is a new era. The old climate statistics may not apply in this new situation. A lot of conventional wisdom has expired.

. . .

I have been laying out the separate strands of the new abrupt climate surge era, but without showing what ties them together. The time has come to see if they make a coherent story. Might there be a common cause for the Big Five extreme weather types?

Yes. Kinks in the jet stream's standing wave can stall eastward movement of highs and lows so that they create extreme weather[25].

And behind the **kinks**? The **slowed polar jet stream** is thought to be caused by **accelerated Arctic warming**, behind which is the **global warming** caused by the 50% excess of **carbon dioxide** (CO_2) overhead. The Arctic is overheating twice as fast as nearby continents because summer sunlight melts floating ice that had been reflecting sunlight, and so the exposed dark ocean surface replacing it absorbs far more heat the next day, melting additional nearby ice. Etc.

Here we see the jet stream reversal concentrated over Kansas, producing violent weather there as a result of slow-moving loops over the Atlantic Ocean.

Elongated *northern* turnarounds also occur, thawing the North Pole in the middle of the 24/7 dark Arctic winter. Another paradox, but with an underlying mechanism clear from the start.

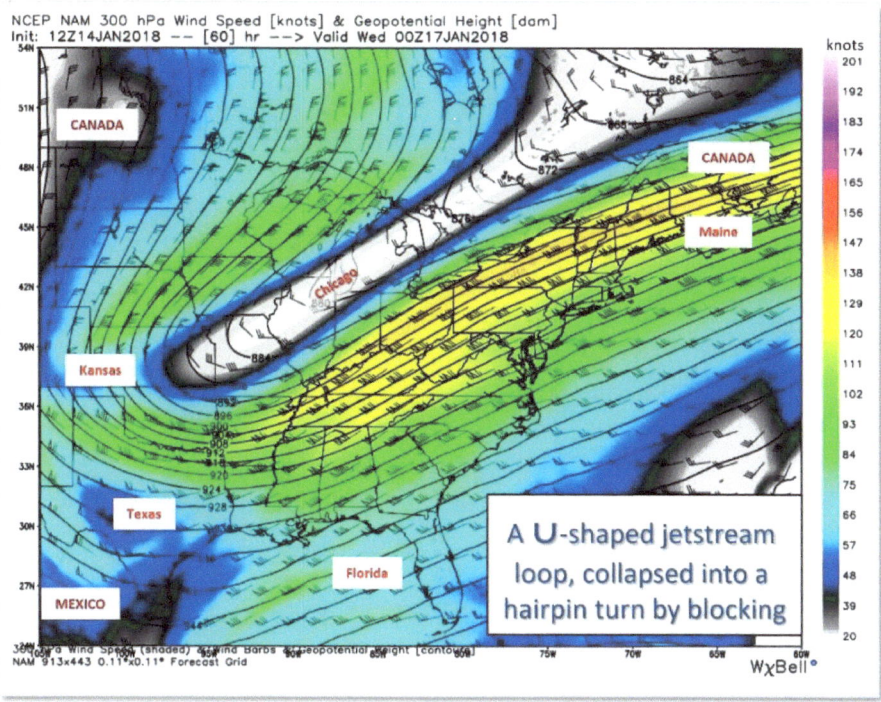

Figure 77. Hairpin turn in polar jet stream reaching from Montreal past Kansas City, and back.

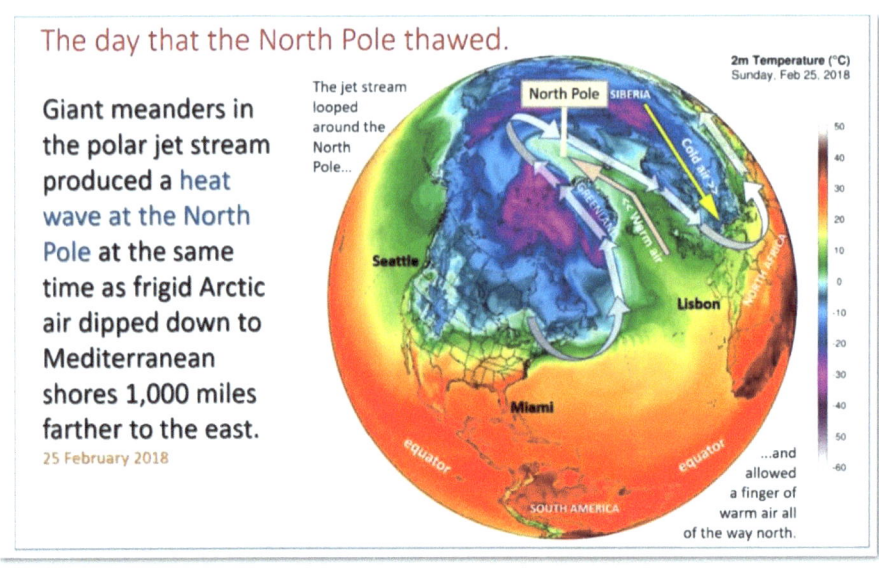

Figure 78. North pole thaw in winter. In February, it is mostly dark 24/7. Yet this warm finger managed to raise the temperature above freezing.

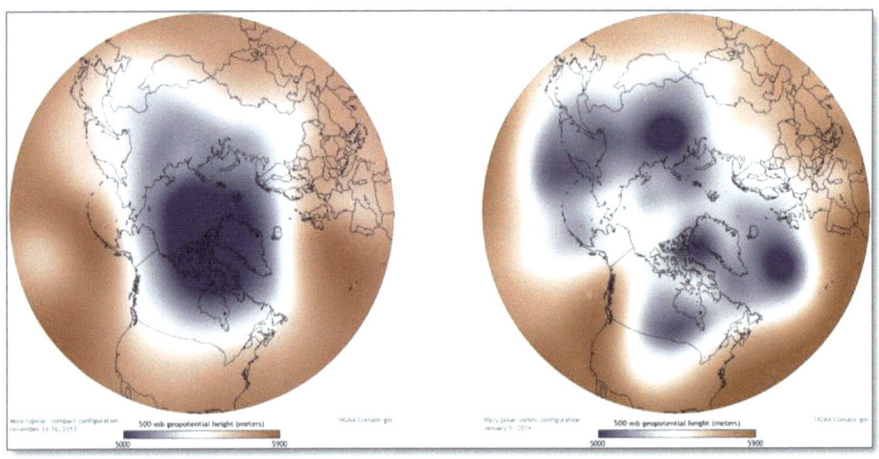

Figure 79. High pressure cell over North Pole (left), whose counterclockwise winds are known as the Polar Vortex, splits into five (on right), each with counterclockwise air circulation. Warm intrusions, as seen in Figure 78, are a consequence. Credit: NOAA.

These Arctic intrusions of warm air likely play some role in reducing the winter buildup of Arctic ice.

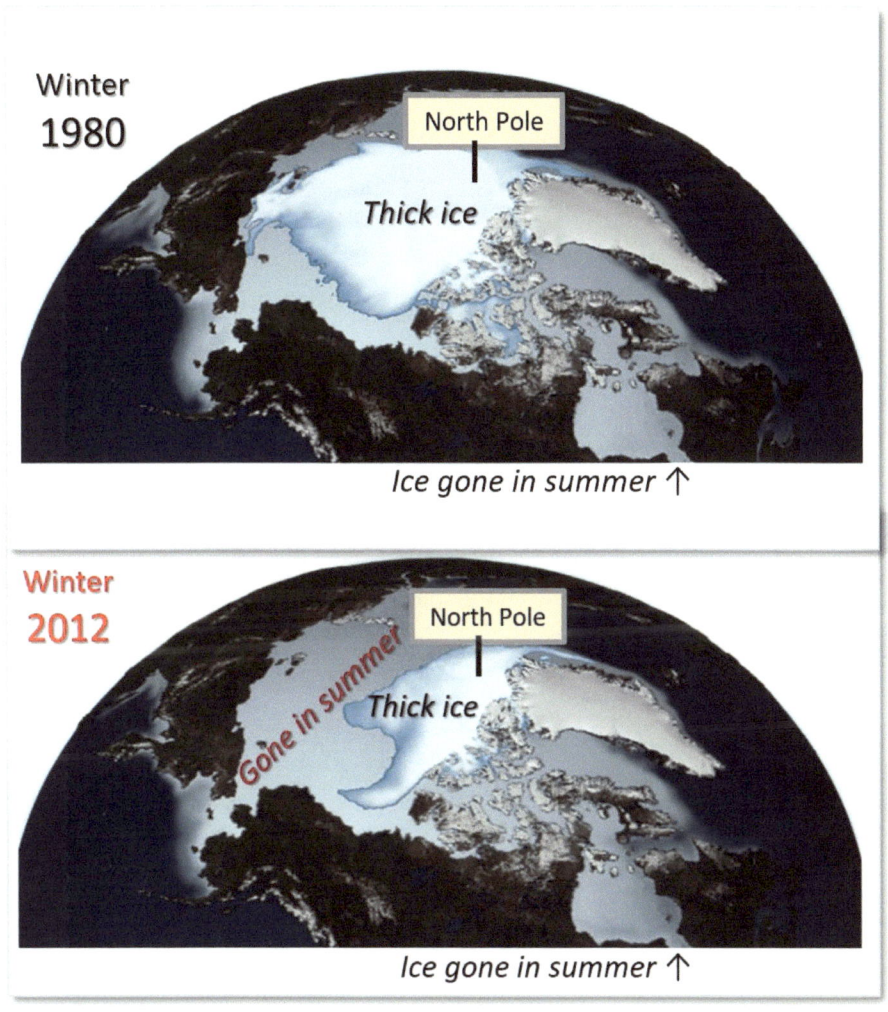

Figure 80. NASA 1980-2012 Arctic loss of multiyear ice. The flat **silver** is where annual ice forms in the winter and melts in the summer. Shown in shaded **blue-white** is the thickness of the multi-year ice. If the 2012 multi-year ice extent is superimposed on the 1980, it shows that almost half has now been lost, mostly after 1998. That lost sea ice is where summer sunlight was formerly reflected back out into space; now the dark ocean can absorb most of that sunlight. And melt back the edges of even more multi-year ice. Thanks to NASA.

Figure 81. Death Valley road sign.

PROGNOSIS
(if untreated)

We now have a *climate emergency*—and the climate community needs to start thinking like the emergency medicine pros, who regularly reevaluate the patient's diagnosis in case new threats have appeared. They act quickly to head off the shock and organ failure that can kill a patient overnight.

One must distinguish between short-term urgent treatments, such as draining a dental abscess, and the longer-term measures, such as cleaning the teeth and avoiding sweets. The climate community now needs to distinguish as well, especially in public statements. Our climate situation is beginning to look a lot like a dental abscess being treated only with fewer soft drinks— rather than with surgery and antibiotics.

It is now time to drain that abscess. Emissions reduction will not do the job that now needs doing.

From Creep to Leap

Climate Creeps characterized Phase One of climate disease, but we are already into Phase Two. Creeps continue, but Phase Two is characterized by those *Climate Leaps* that took place between 2000 and 2012. It's time to explore those extreme weather surges of the 21st century and the jet stream loopiness that seems to be their major cause.

Ramping up suggests a climate creep mechanism such as CO_2 accumulation, somewhere back in the chain of causation. The frog parable is a well-known caution about ramps that seems relevant here, given the current fixation on global average temperature and a merely preventive action strategy, emissions reduction.

One has to be a little careful when reasoning about complex systems which lack an intelligent designer. For example, getting enough oxygen O_2 is an imperative of staying alive. You'd expect our bodies would sense the oxygen saturation in the blood (what is sensed by those little $15 "pulse ox" medical devices that clip to the end of a finger) and raise an alarm if O_2 starts dropping. But as the early generation of pilots discovered when flying up where oxygen thinned out, no such alarm happens even when cognitive performance deteriorates.

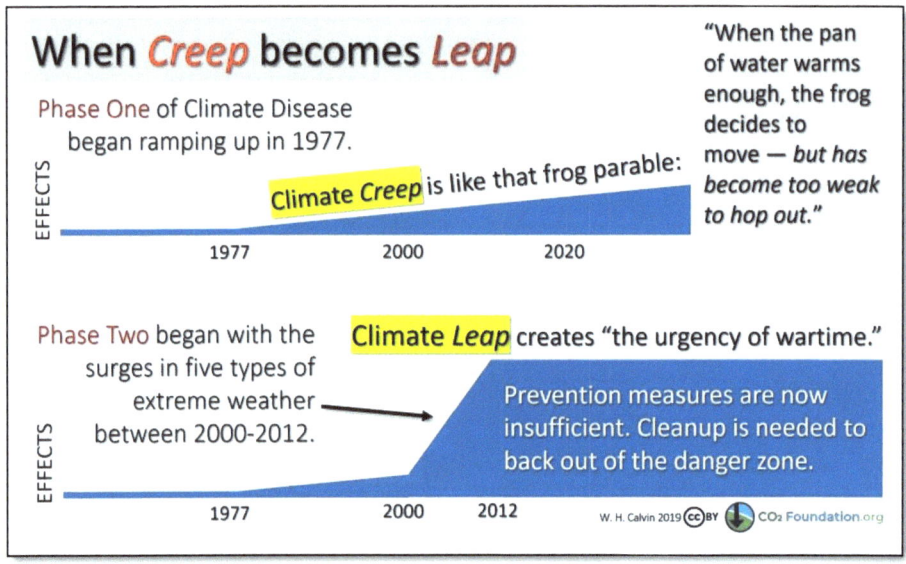

Figure 82. From creep to leap.

It turns out that the body rings the anoxia alarm not by monitoring oxygen levels but by measuring the rise of the CO_2 in the blood. Usually the CO_2 rises and the O_2 falls at the same time as the breathed air becomes stale; it is elevated CO_2 that causes you to breathe faster, not the lowered oxygen—and in evolution that usually sufficed, until balloons and airplanes made quick ascents possible.

So, as anyone who has contemplated evolution's "good-enough" shortcuts knows by now, what is actually monitored to do the regulation of a complex system may not be what a good engineer (an "intelligent designer") would have done.

I say all this because, in the history of climate science, *average temperature* is merely a bookkeeping concept for tracking any imbalance in the heat budget—one which turns out not to track climate problems very well. It was never designed to do so. Climate scientists simply used it as a proxy, figuring that as this 'index' went up, so would climate trouble. But, no. We all try out simple cause-and-effect reasoning before exploring further. A reasonable assumption, but life turned out to be more complicated.

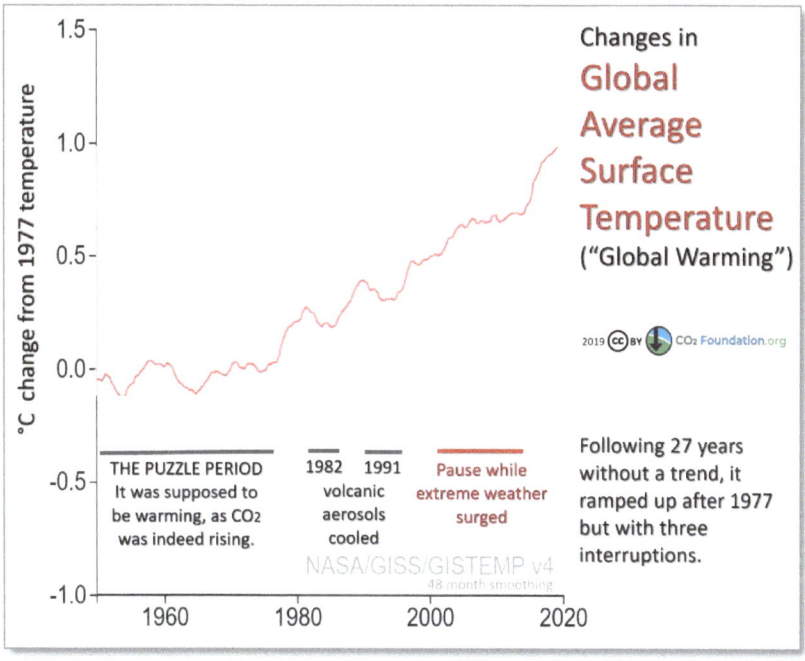

Figure 83. That pause between 2000 and 2012 is best seen with a four-year sliding average.

Back in the days when global surface warming seemed to have paused (in Fig. 84, the "faux pause" from about 2000 through 2012), it occasioned many "Global warming has stopped!" comments in certain editorial pages and contrarian blogs. But that period is also when five types of extreme weather surged. Climate problems increased but global average surface temperature did not. What happened?

Even before the first computerized climate model in 1968 (Manabe and Wetherald told us that the Arctic was going to warm about twice as fast as the earth's surface in general), climate scientists warned of more extreme weather—simply on thermodynamic grounds, what with all of that additional heat and moisture in the air to fuel extreme events. But there were no grounds for forecasting a big *surge* in extreme weather that would latch up into a "new normal," what I have been talking about

83

here. We all thought that a gradual increase in extreme weather statistics would occur, something like a dimmer switch ramping up the brightness.

The way things played out was closer to a traditional light switch being flipped, latching us up into a new mode of climate operation. That's how Phase Two of climate disease presented. By the time of the second mega heatwave in 2010, extreme weather had become our biggest problem, not the next fractional degree of rise in global surface temperature which we persisted in emphasizing. It has taken another decade to see the latch up, what makes it an abrupt climate shift.

Sea level rise and inundation extent

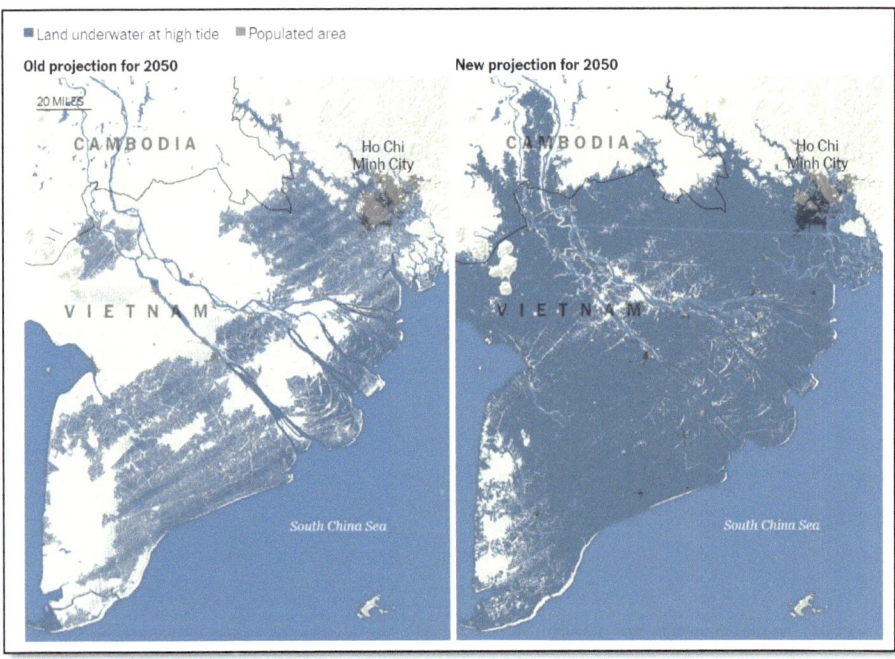

Figure 84. Saigon and the Mekong Delta high-tide inundation in 2050. The old sea-level inundation map for the Mekong delta is at **left**; the new one at **right** with better ground-level elevation estimates shows triple the area inundated. (Thanks to the New York Times for their remake of the scientists' map, 29 October 2019).

Globally, sea level rises for two main reasons[26], both qualifying as gradual creeps (so far). Ocean surface water expands as it warms, taking up more space in its bowl. And hotter summers melt some land-based ice, just as they did when the last ice age ended, and so the increased river flows also raise sea level (in the same manner as annual emissions raise the CO_2 accumulation in the air).

And how fast is this sea level projected to rise? Much faster. According to the 2017 National Climate Assessment of the United States, it is very likely that sea level will rise between 30 and 130 cm (1.0–4.3 feet) during the 21st century.

Back at the time of the 2007 report of the United Nations' Intergovernmental Panel on Climate Change (IPCC), sea level rise was thought to be an end-of-the-century problem, perhaps up 55 cm, and more than a third of that by 2050. Now that is looking like a low-ball estimate.

But that isn't the only problem. We didn't realize how far inland that sea level rise would reach in those developing countries that have not spent the money to do good contour maps of their low coastal plains. They turn out to be much more vulnerable than the IPCC had thought. The threat from 2050 sea level rise just tripled for some major coastal cities because of a new analysis for how far inland their coastal plains will be flooded at high tide[27].

. . .

Earlier, elevation estimates for underdeveloped countries came from NASA satellite radar pings that often measured elevation from the tops of trees or buildings. The newer techniques are more likely to read out the true ground level, where water will be able to flow at the 2050 high tides. That triples the land inundated.

Southern Vietnam could all but disappear; the bottom part of the country on the Mekong Delta will be underwater at high tide. More than 20 million people in Vietnam, almost one-quarter of the population, live on land that will be regularly inundated with saltwater. Much of Ho Chi Minh City (ex-Saigon), the nation's economic center, will disappear with it.

The prognosis? Soon, it will be impossible to buy flood insurance. Or even homeowners' insurance. (Insurance works for random events, not sure things; my father was the executive VP of an insurance company, the guy who evaluated the risks and priced the premiums. I grew up having risky situations pointed out to me as we drove around Kansas City.)

With no insurance, no more mortgages—and so one cannot sell and move. Then no road maintenance; potholes grow into craters. Then lawlessness, the gang warfare problem that comes with partly abandoned regions. People either flee or die in place.

The new inundation assessment shows a similar story for Shanghai, Mumbai, Basra, and Alexandria. In Thailand, more than 10 percent of citizens now live on land that is likely to be inundated by 2050, compared with just 1 percent according to the old roofs-and-trees elevation measuring results; the political and commercial capital, Bangkok, is particularly imperiled.

This new estimate does not include storm surges where high winds push saltwater further inland.

The inundation estimates for developed countries do not change because we created high-resolution digital elevation maps earlier, checked against the traditional surveying methods, and used them to estimate inundation extent.

That's an enormous refugee problem developing soon, not to mention the economic threat as the major financial centers flood and farmers can no longer get loans to buy seeds. And that threat is just from climate creep; storm surge and the new extreme weather also threaten those well-populated coastlines[28].

But the prognosis—those results of inundation— is the same for most of the U.S. Gulf Coast and the eastern seaboard. Except for river flood plains, the West Coast usually has bluffs behind the beaches. Where road cuts have been made in the natural flood barriers, they will need to be filled back in.

Expansion of the tropics and the westerly winds

Pushing deserts farther from the equator changes the latitude at which the westerly winds begin, which is at the farthest extent of the Hadley cell[29].

Figure 85. Senior climate scientists **Stefan Rahmstorf**, **Michael Mann**, and **Richard Alley** at the American Geophysical Union meeting in 2017.

We speak of wet winter and dry summer on a western coastline as a "Mediterranean climate." It happens because the border of the Hadley Cell, where dry air descends, is usually along the southern shores of the Med during the winter. And that is where the first westerly winds develop to bring ocean humidity ashore, down south of Casablanca.

Winter rain. And because those westerlies can 'refuel' their moisture over the Med, even the Middle East gets some winter rain.

During the northern summer, however, the Hadley Cell expands and so the westerly winds move north with them. Summer sun in the Med, because the Iberian Peninsula gets the ocean moisture.

The other examples of a Mediterranean climate (San Diego, Santiago, Cape Town, Perth) lack the refueling afforded by the continent-wide east-west Med. They have rain shadows after the first hills.

The second big climate creep is the expansion of the tropics. About fifteen years ago, a number of studies[30] established that an expansion of the tropics had been going on for decades, pushing the boundary of the Sahara and similar subtropical deserts farther away from the equator. In both summer and winter.

It is currently about 32° from the equator, which is the southern border of California (San Diego is already drying) and of the mouth of the Nile. In the Southern Hemisphere, 32° goes near Cape Town, Perth and Sydney; in South America, Santiago and Buenos Aires.

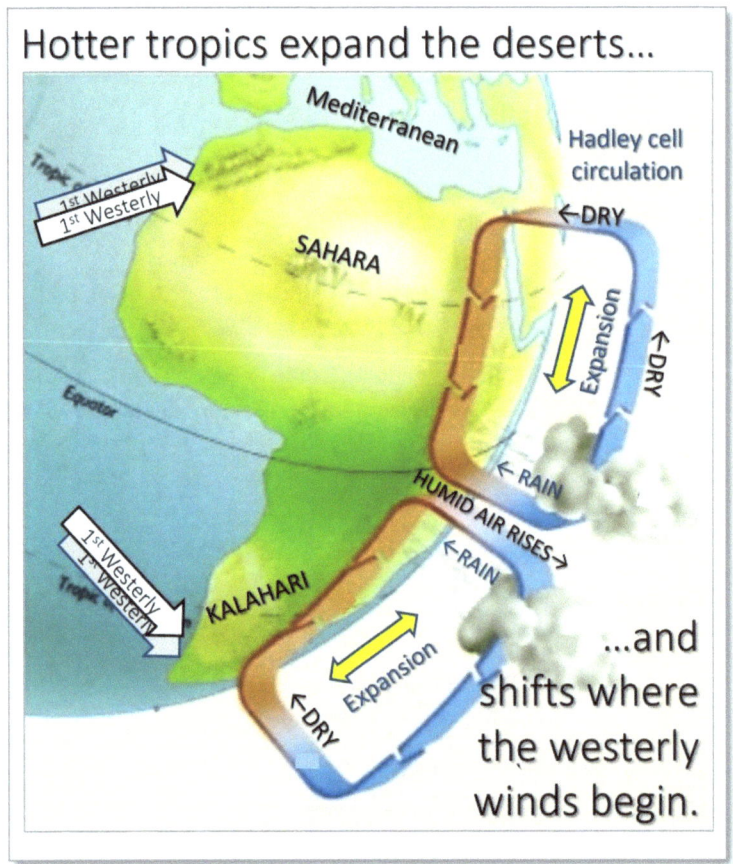

Figure 86. Expansion of the tropics.

All of those big cities are now in the expansion path for the moving deserts. These are all slow shifts, driven by the creep up in tropical temperature, as is the shift in habitat zone for many plants and animals.

The Common Cause

While there may be multiple causes of each, all five 21st-century surges in extreme weather appear to share one common cause: those long kinks of the polar jet stream that cause weather systems to stall in their usual snake-like drift eastward. Narrow hairpin turns form in the standing wave behind a blocking high as the system continues pushing eastward, causing severe windstorms at the tight 180-degree hairpin turn.

When a high-pressure system lingers to become a "blocking high," it gets hotter under its cloudless sky. This creates a kink in the jet stream path to the west of the block—and, with sufficient humidity, that means more derechos, too.

The kinks are also a cause of some of the less-than-triple types of extreme weather, such as seasonal droughts, the out-of-season ice storms, and other unseasonable weather that kills crops.

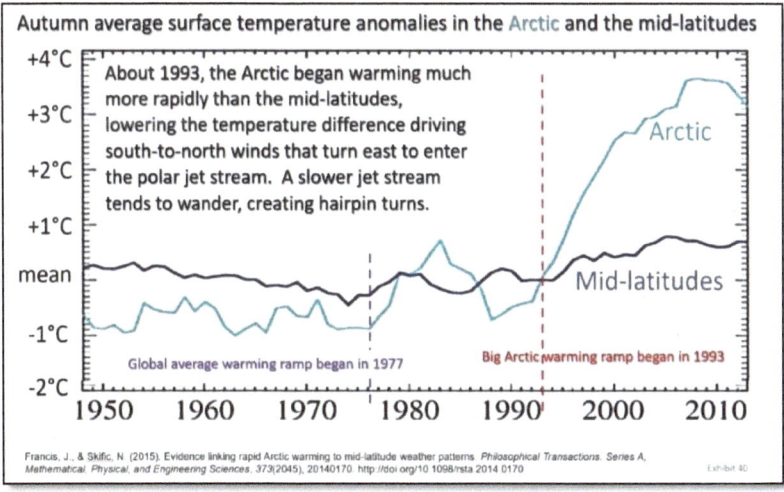

Figure 87. Arctic warming

You can think of the jet-stream loops as being caused by their 100 mph (160 km/h) winds aloft slowing down [31], just as rivers form meanders when they slow down upon reaching a gentler slope but straighten out again upon reaching a steeper gradient again[32]. Here the slowing is because the high Arctic is warming more rapidly than anywhere else on our planet[33].

89

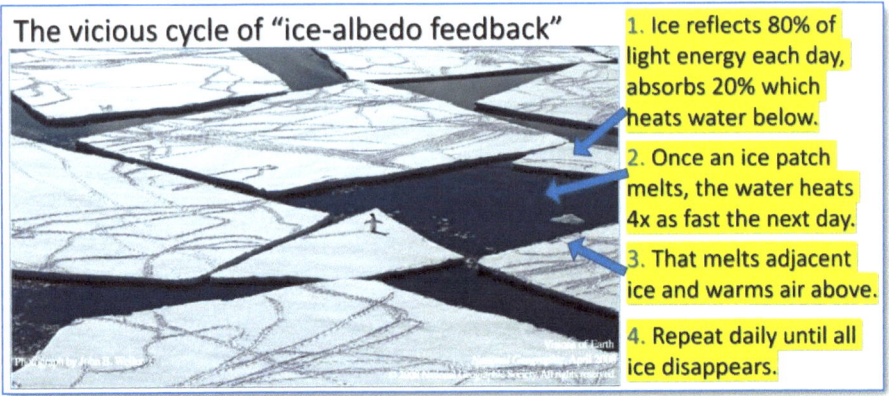

Figure 88. Ice-albedo feedback shows the difference that color makes. Thanks to National Geographic for their penguin tracks near Antarctica.

The surges in extreme weather tell us that we have to move quickly to remove the excess 50% of carbon dioxide (CO_2) from the air; emissions reduction will not help in time, though it is still essential for keeping the climate problems from returning, once we have restored our parents' climate. Yet, despite our efforts, the CO_2 is rising 62% faster than in the late 20th century. Alone, emissions reductions will not get us to safely.

The southerly loops, especially when they narrow to hairpins, are a major source[34] of eight types of extreme weather. I have starred* the five that surged past triple in either annual numbers or severity.

- severe windstorms*
- deluge flooding*
- mega heatwaves*
- stalled hurricanes*
- springtime droughts
- unseasonable freezes
- unseasonable rains
- "fire weather" (hot-dry-windy)*

I have annotated this jet stream snapshot from March 2018 with teaching examples.

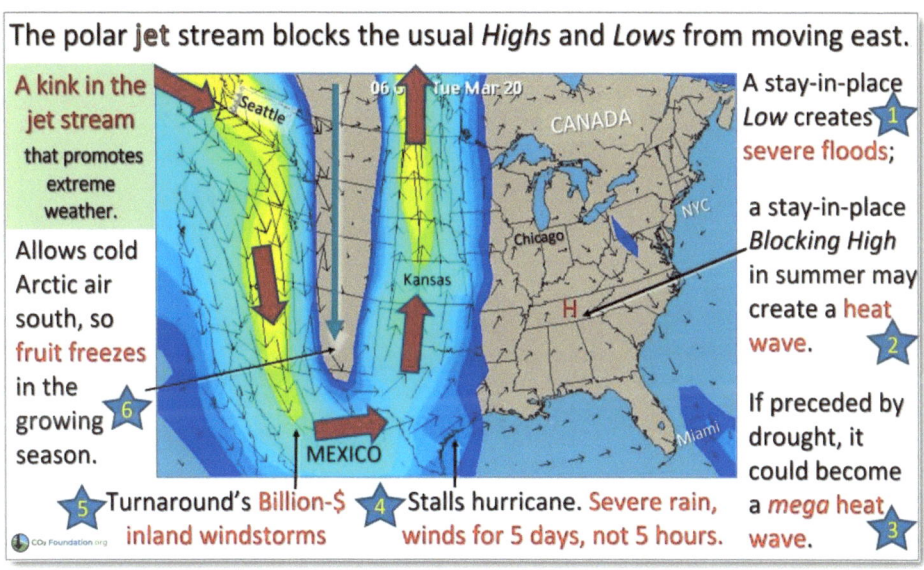

Figure 89. Hairpin turn illustrating the setup for six types of extreme weather. (They all did not happen on the same hairpin!)

Figure 90. In addition to *reducing* global emissions, *remove* the excess CO2.

WHAT TO DO ABOUT IT, *NOW*

In rethinking extreme weather, one must distinguish between the various levels of causation. Burning coal and the kinks in the jet stream are both causes of the new extreme weather, but each operates at a different level of causation.

Think for the moment of an analogous problem with a string of 'causes': On the proverbial dark-and-story night, water starts dripping onto your dinner table. Then the hot overhead lights begin popping out, showering you with glass shards.

No, the upstairs bathtub did not overflow. Inspection of the attic reveals that the roof is leaking. The opening in turn was created by wood rot, caused by insects feeding over the years. Elsewhere the roof shows insect damage, even though not yet leaking.

Chain of causation for extreme weather surges

What to do? Different things on various time scales. Searching for a can of bug spray is not the first thing to do. What you need is a big bucket in the attic, immediately. Tomorrow, call the insurance company; they will send out a contractor to rig a tarp to cover that section of roof. Getting a roofer to repair the roof will take many months on the waiting list.

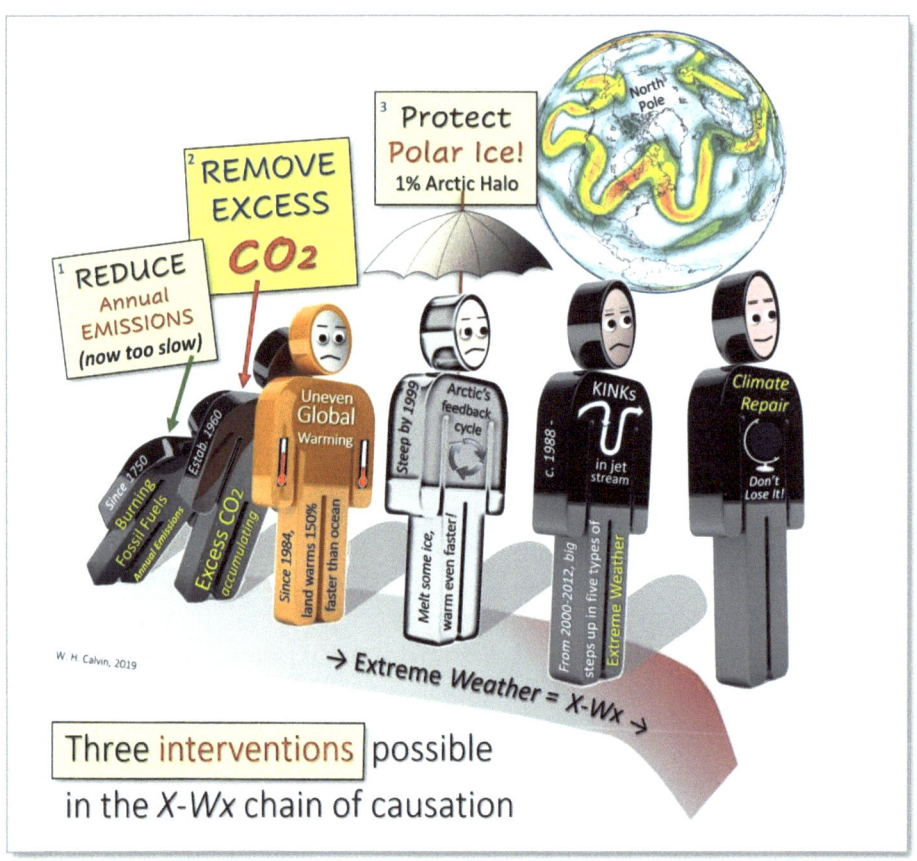

Figure 91. The chain of causation for extreme weather and three points where intervention is possible.

Back to climate's extreme weather surge. Mr. Kinks has a chain of causation behind it: most immediately, kinks in the jet stream that create the extreme weather, caused in turn by ice-albedo feedback that serves to amplify the global overheating from excess CO_2, caused by burning fossil fuels, making cement, and agriculture. In medicine, infections usually have a similarly long chain of intermediate causes, and each may afford a different treatment opportunity.

"Getting at the root cause" is not the only way, nor even the first thing to do, to address the issue. Imagine a dentist who, upon examining your painful dental abscess, offered only the suggestion to drink less soda pop (the root cause) and sent you home. Yet that is all that our leaders have been offering for serious climate action.

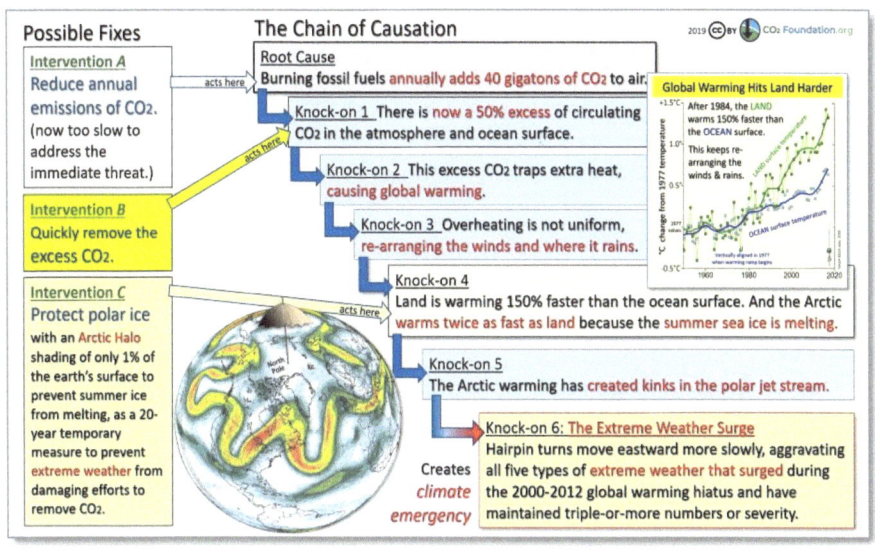

Figure 92. A more detailed list of the chain of causation and of possible fixes.

> *Unlike, say, a global pandemic, sea-level rise is not a direct threat to human survival. Early humans had no problem adapting to rising seas—they just moved to higher ground. But in the modern world, that's not so easy. There's a terrible irony in the fact that it's the very infrastructure of the Fossil Fuel Age—the housing and office developments on the coasts, the roads, the railroads, the tunnels, the airports—that makes us most vulnerable.*
>
> —Jeff Goodell, *The Water Will Come*[35]

"We're the first generation to feel the sting of climate change. And we're the last who can do something about it."

— *Washington State Governor JAY INSLEE, 2019*

"If not now, then when?"
"If not us, then who?"

—*JOHN LEWIS, 1960*

An appraisal of useful climate actions

The efforts of millions of people over the last fifty years has surely done something to slow the steep rise in CO_2.

But not much; it has been overwhelmed by other people using more, mostly to modernize in less industrial countries. I'm not sure we can head that off, given their increasing need for air conditioning to get some sleep on hot nights. We will instead need to counter their emissions for a while.

Getting serious about a climate fix is plagued by some common misconceptions. Most people think that, once we achieve **zero emissions**, the invisible CO_2 will go away as quickly as do the more visible forms of air pollution—and thus we will cool off.

But no. CO_2 does not clear out within several weeks, as do the visible forms of air pollution cleaned out by rain. It takes a thousand years for natural processes to reduce excess CO_2 down to 20 percent above normal[36].

We need to clean up the excess CO_2 ourselves. And complete the job within the next twenty years, *before extreme weather leaves us too battered to act effectively*.

No amount of doubling down on **emissions reduction** is going to do the job in time. To repeat the implications for emissions reduction:

- it is no longer appropriate to call emissions reduction a "climate solution," as I see in headlines or pull quotes almost every day. It misleads people about the serious climate interventions. Slowing is not a solution.

- Even if we achieved zero annual emissions this year, the accumulation of excess CO_2 would still be there, and it is what causes our climate troubles.

- Natural processes take a thousand years to do a CO2 cleanup. That means emissions reduction is now too slow, given the surges in extreme weather.

Small symbolic steps

It is said that most people who feel concern about something only undertake small symbolic steps, if any. Converting them into concerned citizens has been difficult.

Early in 2019, after the flurry of news reports on major scientific reports stressing the need to clean up the excess CO2, the *Washington Post* editors asked activists, politicians, and researchers to list some climate policy ideas that offer hope.

Here is what they got:

1. Open electric markets to competition.
2. Be smart about your air conditioner.
3. Make it easier to live without cars.
4. Prevent wasted food — the right way.
5. Adopt a carbon tax.

Perhaps the *Washington Post* just asked the wrong activists, politicians, and researchers but *there was not a single mention of removing CO2* or other big projects, despite all the news coverage of the 2018 reports that stressed the need for a cleanup.

"Every little bit counts" reasoning has been exploited to add climate's urgency to that of other distantly related worthwhile endeavors that need publicity—say, adding a climate mention to a campaign to turn off the lights when leaving a room. We need a higher standard for claiming "climate solution" because many people are being misled by all of the overclaiming.

We need to convince governments that much stronger climate medicine is now needed. **Size** matters. Now, **quick** matters as well. **Sure to work** has become essential, as we may not get a second chance.

Unless supplemented by something far more effective, emissions reduction efforts will be too little, too late.

. . .

Global warming is no longer a future problem, nor is emissions reduction its most effective treatment. An additional approach to climate relief is now needed, one that can produce bigger and faster results.

Our current approach is to reduce the carbon dioxide (CO_2) emitted from tailpipes and smokestacks. They add to the excess carbon dioxide already in the air, which warms up things to change the climate. But doubling down on clean energy, while still a good idea for the long run, is not likely to improve things in the next few decades.

And zero emissions globally are impossible because the developing countries are going to burn their local fossil fuels trying to modernize—and as warming continues, they are going to need much more electricity for air-conditioning.

But even without that factor, emissions reduction as our primary focus never made much sense. Let us boldly assume that *half* of the annual emissions come from developing countries and that *half* comes from countries that can eliminate their fossil fuels tomorrow. How fast does the carbon dioxide in the air then decline if nature is left to do the cleanup of the excess carbon dioxide (CO_2) overhead?

About one-sixth (half of one-third) of the 140 parts per million excess, or 23 ppm, might be gone by mid-century. That's not very much, merely the amount that we've added in the last ten years and, at the current rate, it will only take seven years to add another 23 ppm. Yet this 23 ppm would appear to be the most we can expect to accomplish from emissions reduction in the countries that could do it. *Yet our leaders keep acting as if that was a sufficient goal for climate action.*

How much might the developing countries increase their consumption, because of their need for more air conditioning to help survive a series of hot nights? Probably much more than 23 ppm.

Convinced, yet? This is important because we have been "betting the farm" on a strategy that won't work.

. . .

99

Major emissions reduction—what scientists recommended fifty years ago and each year thereafter—has recently become the equivalent of putting a band-aid on a cut that now requires a dozen stitches. Yes, our climate band-aid is still worthwhile—I have driven an all-electric car with a "Zero Emissions" license plate since 2013— but that approach has become insufficient to the immediate need. It's what you do *after* you have spoken firmly to your legislators about cleaning up the excess CO_2 in circulation.

Still, let us suppose that emissions reduction accomplishes everything we once hoped for, and does it in twenty years. That still would not get us out of the new climate danger zone fast enough. Zero emissions would still leave the present accumulation for nature to clean up over the years. All of our enhanced extreme weather would continue.

How fast does nature's clean-up happen? Most people assume that carbon dioxide goes away as fast as the visible air pollution does[37]—the time until the next good rain. Actually, it takes a thousand years for nature to remove 80 percent of the excess, though nature should be able to reduce the excess by about one-third before mid-century, mostly by sinking CO_2 into the ocean depths. But after that, removal slows down (we say such a decay curve has a "long tail").

. . .

In recent years, a number of commentators have despaired, saying— usually from economic arguments rather than the scientific ones I just mentioned—that emissions reduction was not going to succeed. They then conclude that we are going to be in terrible climate trouble because of our failure.

But that conclusion does not follow—that's because such commentators seem to think that emissions reduction is the only game in town. There has been a monoculture of ideas, at least in the public discussion. Carbon dioxide removal is seldom mentioned; this omission of a cleanup is not surprising, given their information sources.

Emissions reduction has been the only remedy mentioned, even on such action-oriented websites as *ClimateReality.org* and *350.org*—and indeed in all the pre-2018 IPCC climate reports, from which they have been taking their lead. Governments have not been spending much money on carbon dioxide removal (or, as the industrial, land-based versions of it are sometimes called[38], "negative emissions technologies" or NETs).

Several dozen carbon dioxide removal (CDR, both the NETs and the ocean-based enhancements of natural CO_2-sinking processes) techniques have been seriously proposed[39] but there has been little money for tryouts. I will cover simplified versions of several of them to illustrate the design criteria we need.

Doubling the forests

Though outdated now, the reforestation proposed[40] in 1977 by the physicist Freeman Dyson is the easiest to describe; I'll use modern numbers.

If we doubled forest acreage, taking the global total back to where it was 8,000 years ago before agriculture began to encroach, it would remove 80% of the excess carbon dioxide from our air and make trees out of it.

But don't cheer yet because, like all of the proposals, reforestation has some limitations and side effects to consider first.

Growing a forest to maturity—until it reaches the stage where as much wood rots each year as new wood is produced—takes about fifty years. After that, there is no more net removal of CO_2 because growth is cancelled by decay; one has to cut the forest down, replant, and then protect the harvested lumber for many centuries from burning or rotting, perhaps by sinking it to the anoxic ocean floor. If the forest burns down while growing, all of the captured carbon dioxide goes right back into the air.

Re-forestation would take away acreage from agriculture; hungry people in a future decade would likely cut down the new forest to grow food again. *Not a good plan*, even though it might have worked back when Freeman Dyson suggested it.

Protecting the new forests from fire would require fresh water from somewhere for much of the year. California, for example, is already short on water and diverting it to keeping forests wet during the March-to-November fire season is unthinkable.

And don't forget the equivalent of a mountain pine beetle infestation that kills trees by fracturing their bark following a warm winter, the major cause of those big forest fires in Canada.

I could add the new forest's vulnerability to the surge in fire weather after 2000, but you get the idea: besides being only half as quick as needed, re-forestation is not a *secure* way of doing the job.

"Unable to remove enough CO_2 in the next twenty years" serves to eliminate most of the serious proposals as well—and all of those me-too "climate solutions" in the press release headlines whose contribution would be less than one percent.

That is the current state of affairs, in my judgment, and why we must declare a *climate emergency*.

The time scale of the climate emergency

We have several decades to complete a cleanup—if we are lucky. Here is why: the five big extreme weather surges that surprised scientists—the triple-or-worse windstorms and floods, mega heat waves, stalled hurricanes, and "fire weather"—are really scary, given what they could do to our food supply and to generating waves of climate refugees that, in turn, generate and spread epidemic disease.

Added atop the coastal refugee problem from climate creep, that suggests that we cannot wait until mid-century to *finish* fixing the climate. More such climate surprises can create a world filled with climate refugees lacking jobs, many housed in a neighboring country that, because of a collapsed economy, increasingly finds it difficult to educate or support them. In some cases, if history is any lesson, genocides will occur—if only from uneven distribution of the remaining food.

But there do remain some ways to avoid this, if we hurry. A carbon dioxide removal project will need to be *big, quick,* and *sure-fire.* Leaving this job to profit-seeking private entrepreneurs, as we have done, is absurd. While they are trying hard to make a difference, why should we trust our future to bankruptcy-prone private projects seeking profitable patents rather than common-sense solutions? That doesn't cover the public interest aspects of our current situation very well.

Large government projects to remove carbon dioxide need to begin immediately, without waiting for cost-sharing treaties to be negotiated. That's because even the biggest removal project will take at least eight years before it begins to cool us off (one has to first cancel the continuing emissions before cooling can begin). It then takes another dozen years for this biggest project to back us partway out of the danger zone.

When to do?

Following decades without a serious mention, a number of the big climate reports in 2018 stressed the need to actually clean up the excess CO_2 in the air.

They reflect the growing scientific consensus that emission reductions on their own will be insufficient to **1)** prevent a climate change calamity and **2)** restore the climate to a state that will support life as we knew it.

- The 2018 **IPCC** report[41] concludes that, to avoid exceeding +1.5°C, we must not only stop all greenhouse gas emissions but also urgently deploy programs and technologies to draw down the carbon dioxide already in the atmosphere.
- The U.S. **National Academies of Sciences** (2018 report[42]) states that technologies that suck CO_2 out of the atmosphere will likely be crucial to meeting global climate goals.
- In a 2019 report, the **World Economic Forum** placed "extreme weather events" and "failure of climate change mitigation and

adaptation" on a par with "weapons of mass destruction" as the world's greatest threats.

- In 2019, the **Sierra Club** [43] stated the situation much more forcefully:

  Action on climate adaptation and carbon dioxide removal must be undertaken immediately… and bold action is essential if we hope to protect and restore our human communities and the natural environment in the future. This work cannot wait for five or ten years, and delay will only make necessary changes harder, less effective, and more expensive.

The world has yet to get serious about preserving human civilization as we know it. Those societies that fail to prepare for catastrophe will be more likely to accept authoritarian rule to cope with it. Big industries with profits to protect tend to support right-wing governments, slowing down quick climate action. Here are the *Calvin Interim Design Criteria*. Add your own.

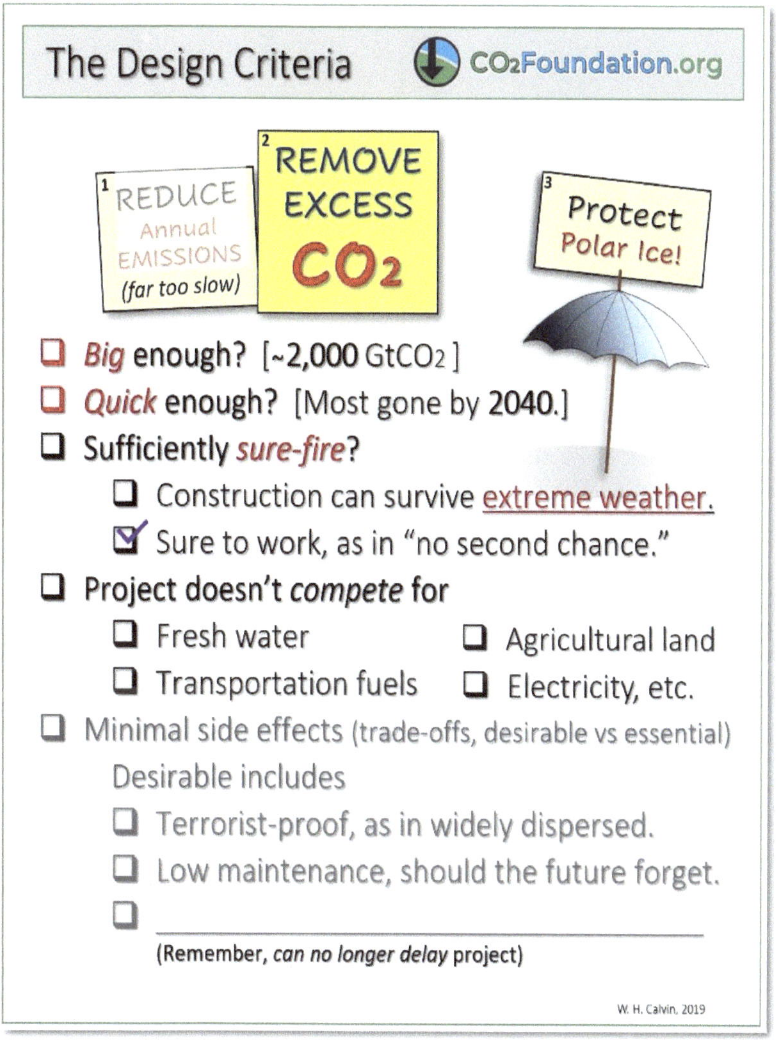

The Design Criteria — CO₂Foundation.org

1 REDUCE Annual EMISSIONS (far too slow)
2 REMOVE EXCESS CO_2
3 Protect Polar Ice!

- ☐ *Big* enough? [~2,000 GtCO2]
- ☐ *Quick* enough? [Most gone by 2040.]
- ☐ Sufficiently *sure-fire*?
    - ☐ Construction can survive <u>extreme weather.</u>
    - ☑ Sure to work, as in "no second chance."
- ☐ Project doesn't *compete* for
    - ☐ Fresh water
    - ☐ Transportation fuels
    - ☐ Agricultural land
    - ☐ Electricity, etc.
- ☐ Minimal side effects (trade-offs, desirable vs essential) Desirable includes
    - ☐ Terrorist-proof, as in widely dispersed.
    - ☐ Low maintenance, should the future forget.
    - ☐ _____

(Remember, *can no longer delay* project)

W. H. Calvin, 2019

Figure 93. *The Calvin (2020) Interim Design Criteria.* For the push-pull pumping example in the Q&A, most boxes are checked.

A paradigm shift to repair climate

Pardon me for my focus on the U.S. in what follows. But read on—other countries will need to do something similar at the same time; otherwise, they will be risking further inaction by the U.S., the world's largest contributor[44] to the excess CO2 accumulation overhead.

Moving the U.S. Congress to act could take years—years we no longer have, having wasted action opportunities for the last fifty years. We need a workaround.

So, to ensure a fast response, consider what we might do to get started without them. Just as California's initiatives provided 20th-century leadership on national-level air pollution efforts, so a *Governors' Design Initiative to Repair Climate* could provide badly needed 21st-century leadership to jump-start climate action.

Our current *analysis* of the climate problem focuses on a global +1.5°C coming up by 2030; on the *action* side, it calls for more emissions reduction (authoritative statements often do not mention anything else, as if one had to achieve zero emissions before starting CO2 removal, another logical fallacy).

I propose a paradigm shift in how we scientists focus the attention of others: use the extreme weather surges for which we have such a threatening track record. For the action side, we need to focus on removing enough CO2 before 2040 to back out of the danger zone. That may also require temporarily dimming the Arctic sunshine to reduce extreme weather that endangers the drawdown projects.

That also gets climate scientists out of the forecast business—a good thing, as they do not yet understand extreme weather trends well enough to forecast future decades, good as they are at forecasting the creeps. While an increase in extreme weather had long been emphasized, I don't think that anyone talked about a sudden surge until they started to happen. Computer models simply were not detailed enough to handle such nonlinear features of climate dynamics as the loopiness of the jet stream. Now we are seeing models[45] that can examine some features such as hot and dry summers associated with loopiness. But predictions are unlikely to be very detailed as the models would run into the chaos theory issues when random events cause detours.

From my neurophysiologist's perspective on the dynamics of another complex system, let me briefly comment on the state of the climate science when it comes to predicting the paths and strengths of future winds and rains. Climate science was not advanced enough to warn of the 21st-century abrupt climate shift represented by those five types of surge in extreme weather. And I don't expect it to be good enough to warn of sudden shifts for some time to come. That's a difficult, maybe unsolvable, problem given the chaos considerations.

We should not be expecting the impossible from the climate scientists, nor waiting for advice from them about how we should proceed. That's not their specialty. The existing evidence is already quite sufficient to tell us what we should be doing.

A Course of Climate Action

Around medical schools, we talk of intervening in the natural course of a disease or disorder. For climate disease, there are three broad categories of climate intervention.

1. **Reducing emissions.** The only course of action known to most people is a "Use Less" approach to prevention that made perfect sense fifty years ago.

Several things have happened in the meantime which tell us that we cannot rely on emissions reduction, logical as it sounds to many people. First, it has not worked on a global scale. The annual bump up in CO_2 concentration has increased 63% for the 21st century years. There is no reason to expect that a low-carbon fuel diet will begin working, given the need of developing countries for air conditioning during hot nights.

It became apparent a decade ago that nature will take a thousand years to clean up after us, not a century or two. The CO_2 problem is not like more familiar forms of air pollution, cleaned up by rainfall.

That leaves us, broadly speaking, with two distinct approaches to our long-term overheating problem, *shade* and *cleanup*.

2. Shade, aka Solar Radiation Management (SRM), tries to bounce back some arriving sunlight and thus cool us off—even though CO_2 continues to build up. It is usually presented as trying to mimic the aerosol haze after a volcano erupts, but we have long been unintentionally practicing such "geoengineering" via air pollution[46].

3. Cleanup, aka Carbon Dioxide Removal (CDR; the land-based-industry versions are sometimes called Negative Emissions Technology, NET) has a number of subcategories, such as Direct Air Capture, Ocean Fertilization, re-forestation, push-pull ocean pipes, etc. There is an excellent evaluation of how to make (and store) dry ice by von Hippel[47]. It's a non-starter but he analyzed the problems which other land-based NET projects will need to consider.

But now there is an urgent short-term problem from the new extreme weather, about which the discussion is only beginning. None of the several dozen CDR proposals[48] adequately addresses the urgency of doing the cleanup job. None of them could remove CO_2 fast enough to remove even one-third of the continuing emissions annually—and thus they will not cool us at all. We have been aiming too low.

In many nonlinear systems, it can be harder to back up than go forward ("hysteresis"), meaning that it could take extra decades to back out of the danger zone—another reason to treat our climate prospects as an emergency.

Climate engineering and two simplified examples

Rather than work through the list, what follows is one idealized CO_2 removal process that shows how it might be done, providing an example of what it will take to satisfy those big-fast-secure criteria, countering continuing emissions by 2027, and then starting the cool-down,

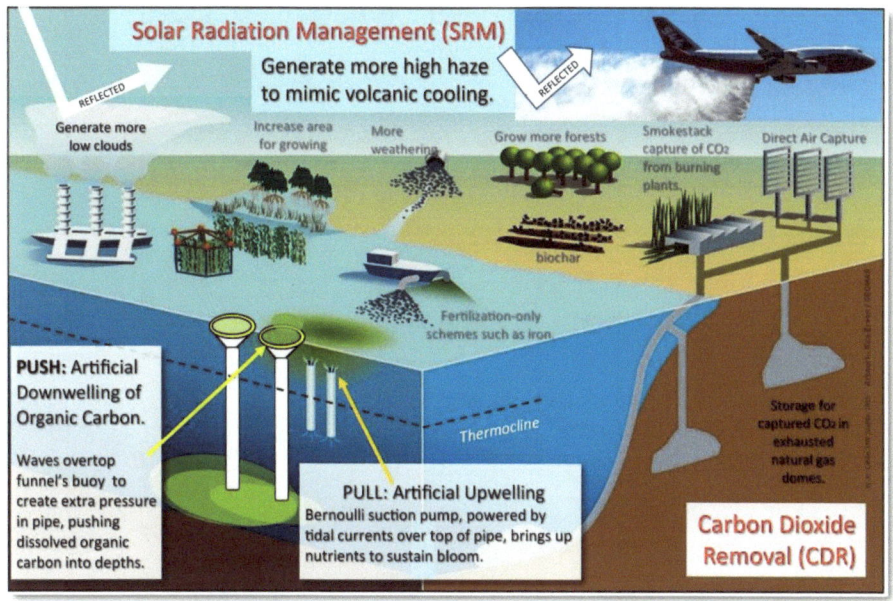

Figure 94. The various proposals for shade and CO2 cleanup.

finishing the job by 2040. It is the biggest and fastest way of proceeding that I can ballpark.

It is intentionally simplified, so assume that a working group of experts will be creating something better, using such design criteria and then prototyping a few candidates over the next four years before a mass deployment begins. I'm not trying to sell a particular solution, only a design *process* that could lead us forward to a workable solution. There are more details in the Q&A at the end of the book.

So, let me describe the path to two **climate interventions** chosen more for their simplicity than known efficacy:

For scale, the aim must be to remove enough excess CO2 by 2040 to back out of the danger zone for hidden tipping points and simultaneous hits from the five extreme weather types that have already surged.

First, we will need some background on how the ocean usually recycles carbon, in order to appreciate how we can augment it to take the excess CO2 out of circulation for a few thousand years.

The ocean's carbon cycle: The in's and out's of CO2

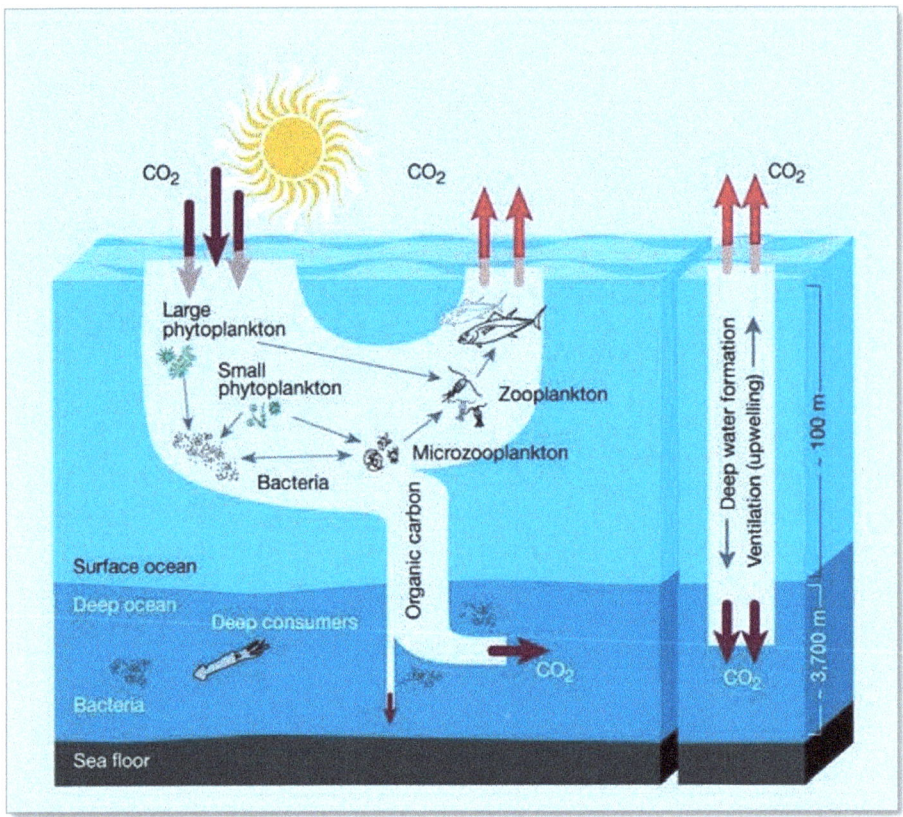

Figure 95. Two routes to the depths: (Left) **settle out**, and (right) **go along with the flow**.

Photosynthesis by algae captures a CO_2 molecule, strips off the O_2, and combines the Cs with each other and with nutrients in the sea water to make an 'organic' carbon molecule such as sugar. Leaves on land do the same thing except they have to first use evaporation to pull some water out of the ground to transport the nutrients up to the site of photosynthesis in the leaf. Globally, about half of our oxygen comes from rain-dependent leaves, half from algae.

But algae and other plankton stop working within a week and so decompose slowly into 1) heavier parts that sink, 2) particles that stay suspended at varying depths, and 3) the dissolved organic carbon (DOC) that surface-layer bacteria eat. Their respiration converts it back into CO_2, and thus it escapes back into the air. Feces also export organic carbon from the surface layer, but most break up on the way down and fragments stay suspended at some depth, becoming nutrients available for constructing new single-cell organisms via cell division.

The deep ocean already contains fifty times more carbon hanging around than all of the excess CO_2 we must take out of circulation. An obvious strategy is to not wait for particles drifting down or for diffusion but rather to sink surface waters into the depths as a flow ("vertical convection"), *the downwelling flow carrying below not only plankton but all of that DOC food for bacteria that will soon release CO_2 back into the air.* This mimic of natural downwelling is not to be confused with injection of industrial CO_2 where it will be carried into deep ocean[49].

Once we have augmented nature's natural downwelling, we can augment nature's natural upwelling by pumping up additional nutrients[50] from just below the thermocline to fertilize growth in the surface layer exposed to sunlight[51]. This tends to bring up all of the essential nutrients, not just iron[52]. [There is much more on this topic in the Q&A., starting at fig. 100.]

THE TREATMENT PLAN

Four years for research and simultaneous prototyping, field trials, etc. (the time frame for a Design Initiative) sounds like a short time but recall how fast things were made to happen during World War Two[53].

The 2020 Manhattan Project

During the Second Industrial Revolution (radio, electrical power, cars, planes) a century ago, Thomas Edison created the cross-disciplinary invention factory[54] where scientists and abstract thinkers work cheek by jowl with machinists and electricians and other hardware tinkerers. We now need a four-year gathering of international experts on climate engineering, solving an urgent problem[55] in an Edison-style "design shop."

My present concept of a Governors' Design Initiative would bypass the initial years of consensus-building in Congress. It would immediately begin four years of design and field testing, getting the effort up to speed before the federal involvement finally arrives, allowing manufacture and deployment to begin in earnest.

Those years are reminiscent of the process during the 1942-1945 Manhattan Project that gathered physics, math, chemistry, and engineering professors of varied nationalities at Los Alamos—that canonical example of a hurry-up design project that ended up with two very different prototypes (gun barrel and implosion), both of which worked.

I have been reminded that, to many people, "Manhattan Project" simply means "Atomic Bomb" but my reference to it is about its *process*, not its product.

Treatment plan for climate disease

Four years for research and simultaneous prototyping, field trials, etc. (the time frame for a Governors' Initiative) sounds like a short time but recall how fast things were made to happen during World War Two[56] (my other reason for sticking with the Manhattan Project analogy).

Manufacture and deployment would ramp up over the following five years. This type of rapid buildup is traditionally assigned to the military; DARPA might be appropriate to lead a Pentagon effort and would certainly wish to be kept informed during the design phase. But how much can they do during the design phase without Congress speeding up?

Time, I suggest, to lean on the tech billionaires via a few governors doing the asking.

Cleanup complete (or at least good enough) in the next twenty years, though CO_2 removal will still be needed afterward, to counter continuing emissions from less developed countries.

Only after 20% of the CDR array is built and working—say, 2027—would we be taking out as much CO_2 each year as continuing emissions were adding. Only after that does the CO_2 accumulation start to drop. Cooling finally begins.

Redoubling emissions reduction efforts might buy us a year's advance to 2026 but, by themselves, reducing emissions will no longer fix our extreme weather problem, even though good for the environment more generally.

SRM does not have quite the lag-time problem as the CDRs, but safe-to-try SRM only applies (in my view) to the 1% *regional* cooling the high Arctic, not to *global* cooling.

Timeline for Defeating Extreme Weather

To scope out this project, let us start by defining CO2 removal (CDR) in a way that avoids big words such as sequestration, or the unneeded distinctions between organic and inorganic carbon.

CDR simply means taking carbon out of the circulation loop—literally. It does not have to be forever. It's the excess CO2 overhead that overheats us. It's the excess CO2 in ocean surface waters that threatens our food supply. How much CO2 do we need to sidetrack? That depends on how fast we want to accomplish the CO2 cleanup.

Scoping a big project can be done without waiting for better numbers. Let us begin with the fastest cleanup that I can imagine, were everything to turn out just right. You might think up something quicker or say that some part of my simplified cleanup cannot be done as fast— but bear with me until we get my fastest possible estimate. I am going to outline the scoping procedure so that you can run your own calculation. Any architect will recognize the procedure.

Unlike present efforts, now we must start with an evaluation of tolerable risk—the risk of waiting until later—what we have been doing for a half-century already. In my scoping evaluation, we need to clean up all of the excess CO2 (about 140 ppm of the present 420 ppm) by the year 2040. (Try, if you like, running the calculation below with a 2030 or 2050 goal; that will give you an idea of why I picked the year 2040 for cleanup complete.) Once this time to completion is established, one then works backward to see how big the project needs to be for achieving that goal in such a time frame.

If 2040 is the goal, then what built capacity would we need so that its annual production (in GtCO2/yr) would achieve enough CO2 removal and storage by 2040? After the completion date and the size is established, we can worry about costs and tradeoffs.

- Suppose that one looks up the amount of CO2 in the air. There is currently a CO2 excess of 140 parts per million (420 minus 280 ppm). We also know that an 8 GtCO2 addition is needed to bump it up 1 ppm, so we decide that we must remove about **1,200 GtCO2** from the air's total of about 3,600 GtCO2. That's removing **-60 GtCO2**/yr (containing about **15 GtC/yr**) for the project period of twenty years. (When ballparking, use rounded-off numbers so as to not imply precision[57].)

- *Too simple*, since there is a constant exchange of CO2 between the air and the ocean surface waters, what is called *equilibration* in chemistry. When one begins removing CO2 from the air, the ocean *surface* waters will begin returning some of their excess CO2 to the circulating air; it was only buffered, what causes the surface ocean acidification. There is some time lag for equilibrating, which I will ignore for the moment.

- There's a carbon reservoir (mostly dissolved organic carbon, soon to become CO2 again) of 1,000 GtC in the *surface* layer of the ocean, of which about **300 GtC** is excess, soon to become **+1,200 GtCO2**. That doubles the needed cleanup. Allowing 20 years for doing this, you'd think that **-120 GtCO2/yr** capacity might do the job.
  Note that, whether we remove the CO2 from the air or by sinking blooms from the ocean surface into the thousand-year depths of the ocean, either way we still clean up the atmosphere's excess CO2.

- But we forgot 20 years of continuing emissions of **+40 GtCO2/yr** (probably more, as it grew by 20% between 2015 and 2019). That's +10 GtC/yr (the conversion factor is really 3.7, not 4, but I'm rounding off).

- Hold on. **-10 GtC/yr** is the approximate rate that natural processes such as downwelling currently sink carbon from surface waters into deep ocean storage (whose reservoir size is about 38,000 GtC), which takes it out of reach of the atmosphere. Put another way, continuing emissions are cancelling out nature's major long-term process for taking excess CO2 out of circulation. So, we are back to needing **-30 GtC/yr**.

- However, we are not ready to go and must allow at least five years to design and prototype. So, divide by 15 years rather than 20 and we are up to needing about **-40 GtC/yr** or **-160 GtCO2/yr**.

- We must also allow for deployment time, for ramping up that capacity rather than starting at full strength. If we do that ramp over five years, it puts us up to needing closer to **-200 GtCO2/yr** operating between 2030 and 2040. Now you can see why there is such a range of estimates for a cleanup.

- Next, allow some extra time for delays and for things going wrong. More experienced project planners, I am told by architect friends, would likely double that **-200 GtCO2/yr** estimate to provide a safety margin, just as they do for strength when building bridges.

Cooling's start would be delayed until after 2027, and its influence on taming extreme weather would likely be delayed into the 2030s. That would seem to be the biggest and the fastest that we can do over the next twenty years, even with a really ambitious project.

Even if we sank all the excess from air and surface ocean (that 600 GtC), it would only increase the dissolved carbon in the depths by 1.6%. The emissions that are already sunk into deep ocean stay sunk for more than 1,000 years (6,000 might be a better estimate). It comes back up very slowly, requiring quite a few passes through the surface ocean before making it back into the atmosphere to again contribute to the insulating blanket. The water circulation time for "ventilating" underestimates the true storage time. From radiocarbon dating[58] of deep water, the slowest fraction takes 12,000 years to become atmospheric CO_2 again via bacterial respiration. That's good news, finally.

Alas, there is at least one more missing actor: the warming effects of the other excess greenhouse gases (GHGs). Cooling things off is, after all, our goal. Rather than separate cleanup projects for each of the excess GHGs, we can just remove some extra CO_2 to counter their heating effects. Methane is gone in several decades, so after the CO_2 excess is done, we will still be removing CO_2 and adjusting the annual amount. But how much more do we need *now*? The quick way of estimating that is from the pie chart of GHG proportions. CO2 is three-quarters of the warming emissions, so a 33% increase in the amount of CO_2 removed should cover that base.

There went our safety margin. Time to scale up again (left as an exercise for the student).

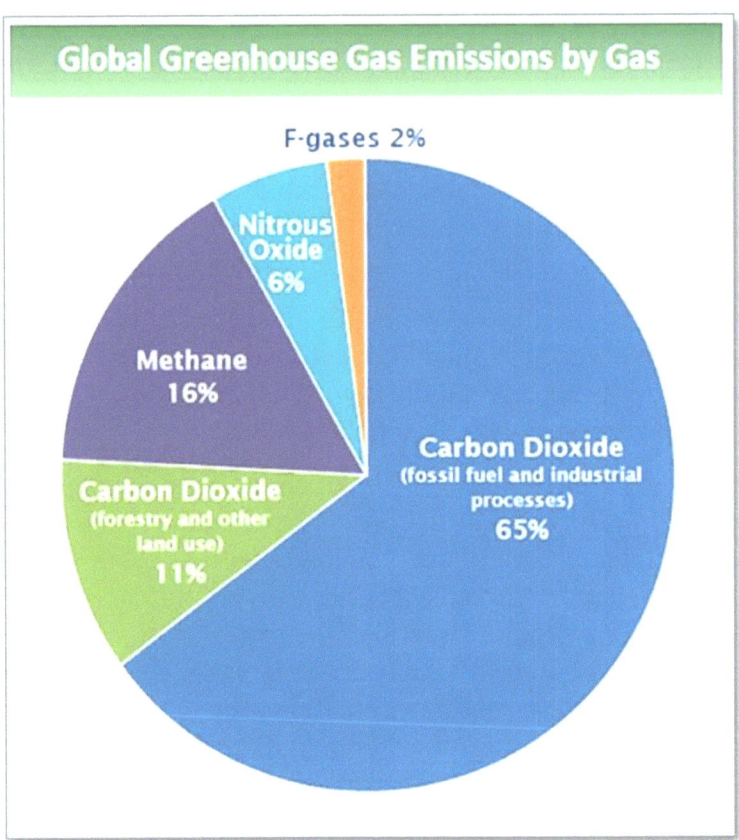

Figure 96. GHG proportions, from IPCC (2014) global emissions/year. Annually taking CO2 out of reservoirs (fossil fuels, forests) and putting it into circulation is about 3/4s of the overheating problem. A 1/3 increase in CO2 removal would take care of the others as well.

. . .

As mentioned earlier, I am trying to show how *fast* a big CO2 cleanup could conceivably be done, allowing you to supply your own judgment about whether we should accept greater risks to order to save money. "Sticker shock" will cause some to propose scaling things back to removing lesser amounts, so here are some data for what that buys.

- If we reduce our goal to removing only 58% of the excess, to meet the original 350 ppm goal of *350.org,* that only gets us to the 1988 levels which had already caused major trouble.

- Removing 75% of the excess gets us back to the CO2 levels of the 1970s before the overheating ramp began.

- Removing all of the excess takes us back to the CO2 levels of about 1750-1800 as the first Industrial Age (steam engine to railroads) got underway. But that 280-ppm level was already at the all-time high for the ice ages, and so not necessarily safe.

Each year of additional delay in getting started could see major economic hits; it also means paying for a bigger plant capacity when we finally begin (try running that calculation for a 200-ppm excess at the start). There might be a poorer economic base for taxation then, because of further disruption by extreme weather or resource wars. Furthermore, the additional battering reduces the probability that our efforts will eventually succeed.

Evaluate any serious climate fix proposal with this scoping approach to see if it is *big* enough, *quick* enough, and also *sure-fire*: able to be protected from terrorists, extreme weather, and economic crashes. There are lots of good ideas around for the long term—that's after we reduce the extreme weather—but few inventors have been paying attention to the more immediate extreme weather threat. Different criteria now apply, and speed is not merely nice but necessary for the climate emergency.

. . .

You might think that removing the global excess of CO2 should be a big international endeavor, like cleaning up the plastic floating in the ocean's eddies, with costs shared according to contributions.

But think how long it would take skilled diplomats to get either cleanup project up to speed. Thanks to a half-century of ignored scientific warnings, there is now no time left in which to do the organizing in the usual way, let alone sort out the cost-sharing. *We have already sped past that exit on the freeway to Hell.* With a fast train coming down the tracks, one should not stand around awaiting a diplomatic reply to the most recent bargaining round of cost-sharing.

The implication: *big industrial countries will each have to independently undertake their own crash program*, welcoming observers and foreign collaborators but not waiting for them.

After another decade, treaties for the long term can fill in behind, with international climate commissioners then drawn from those who have already been participating in getting things done. It's a job for experts, just as the U.S. Congress decided when establishing the independent agencies such as the Federal Reserve Board. (Just try to imagine Congress setting interest rates and bank reserve requirements on the fly—as they used to do.)

The biggest of the recently reviewed "negative emissions" proposals[59] would sink less than **-3 GtCO2/yr**, which contains 0.8 GtC/yr. Therefore, they would not counter the **+40 GtCO2/yr** of continuing emissions, let alone draw down the excess CO2 and eventually cool things off. Size matters.

Two current CDR proposals that seem big enough for the present situation are 1) quickly doubling our forests, and 2) plankton farms that sink both the new living plankton (and the 300-fold greater amounts of dissolved organic carbon) to the deep ocean for thousands of years.

Both proposals are currently viewed as flawed, but a working group of experts should be able to design something better in a few years, such as wind/wave-powered systems that mimic, on a large scale, natural ocean upwelling and downwelling.

Governors' Design Initiative to Repair Climate

Let me conclude the presentation with the "executive summary" of a "Manhattan Project 2.0." Following it are some frequently asked questions and topics with more technical details.

The best indicator of climate change is no longer a fractional-degree change in global warming. Now we have at least *five major shifts in extreme weather* requiring faster action than can be achieved through more effective emissions reduction.

Climate action now must shift from *prevention* to *repair*. We should clean up the excess CO2 by 2040. But, just to protect the cleanup project, another temporary protection project may need to move quickly to address the loopiness of the jet stream, perhaps by shading the Arctic until global cooling kicks in.

Both projects may need four years of design and field trials before deployment. Congress is likely to be slow, but a *Governors' Design Initiative to Repair Climate* could now orchestrate those design-and-prototype years for CO2 removal, largely funded by the many tech billionaires who are already familiar with doing big design projects quickly. The Design Initiative could also build support in Congress, with constituents approvingly pointing to specific designs fresh from the experts.

The current public discussion about the climate crisis only focuses on reducing our yearly CO2 emissions via greener fuels and little personal efficiencies. It's a diet, another "Use Less," and it seems to have much the same success rate as food diets. Emissions reduction is a long-term strategy; it omits such considerations as how civilization will survive while the plan plays out. Such a fossil-fuel diet is insufficient to the task: too little, too late.

. . .

Fix the cause and you fix the problem? Not anymore. Focusing only on emissions reduction is very outdated, suitable only for a half-century back, when scientists first began warning, *every year*, that our only planet could seriously overheat.

The emissions-reduction prescriptions might have worked had they been widely implemented fifty years ago. But the CO_2 overhead has instead increased from 320 ppm to 420 ppm, more than triple the *excess* CO_2 present in 1965 when White House science advisors first considered the threat.

But now the new extreme weather demands a short-term survival strategy in addition to sensible preventative measures such as emissions reduction.

Currently, we are simply slowing our approach to chaos—a little. The current prescriptions are now painfully inadequate because they omit any notion of climate repair, such as a cleanup of the excess CO_2.

We can do better. And, as it turns out, we must *quickly* do much better for two reasons. We are approaching coastal high water much faster than we are re-locating people inland. And because Phase Two of climate disease has already set in, surprising even climate scientists. Now we are being battered by a surge in five types of extreme weather which will increasingly limit our abilities to repair climate.

It might take four years to get the U.S. Congress up to speed. That's four years that we don't have, not anymore. We need to get moving on climate repairs—this year.

We are already being battered by a surge in five types of extreme weather. That escalation in extreme weather is what we are really up against, not just the slow rise in average global temperature and sea level.

Extreme weather surges now create a new urgency and an even larger policy gap. Because of the thousand-year natural storage time of excess CO_2 in the atmosphere, reducing emissions will not suffice.

This new climate instability is why things have become so urgent; it is what makes much of our present climate discourse sound outdated, yesterday's version of the big issues commanding our attention. The 2018 IPCC Special Report says that carbon dioxide removal strategies (CDRs) are now needed.

For the next few years, we need a *Governors' Design Initiative*, with a finance committee of tech billionaires, to design and prototype CO2 removal while national governments get up to speed to handle deployment.

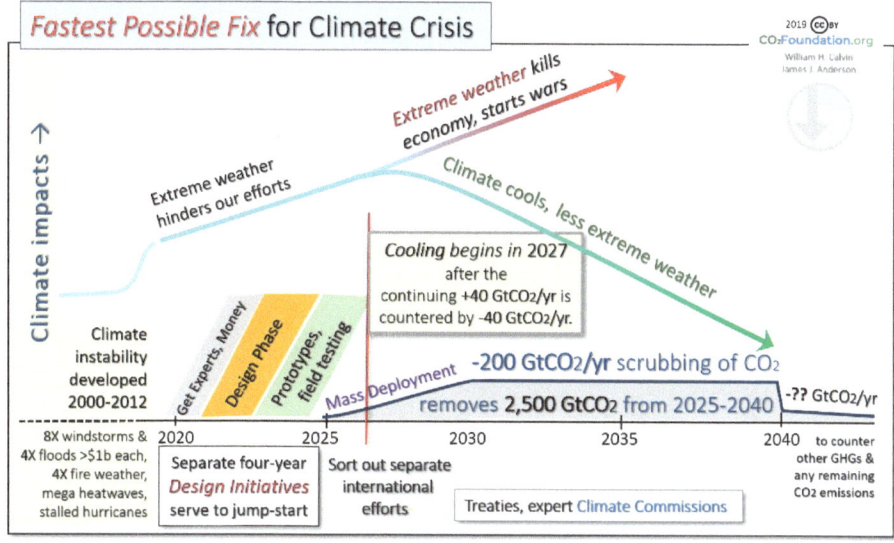

Figure 97. This is the fastest CO2 removal project that I can imagine.

This is what gets us started. Better designs will replace them in another decade, just as U.S. building standards are revisited every three years by experts, so that recent buildings are more earthquake resistant than those built to high standards twenty years ago. So, in judging what climate action shortcomings we can tolerate, remember they might be short-lived. *The need to start big is now urgent.* We now need to take some chances because we were so slow to take overheating seriously.

The Wrap

While it seems like common sense, emissions reduction is a mere band-aid for an injury now requiring a dozen stitches. We need to start with expert planning and prototyping—essentially, a *Manhattan Project 2.0* for taking the excess carbon dioxide accumulation out of the air.

The most effective action you can currently take as an individual is not cancelling your fossil-fueled vacation but rather insisting on a CO_2 cleanup. You repeatedly make your elected officials aware that you expect them to organize a big cleanup. And quickly, before the next election.

Even without another surprise surge in extreme weather, big-and-quick carbon dioxide removal is now our survival strategy. Our situation is now the equivalent of having to prepare for a great war, already looming on the horizon. Clearly, we are starting late but there are still possible ways of fixing the problem—if we treat this as an emergency. There is nothing hopeless about our situation, as some commentators are beginning to suggest—ones who have been led to believe that emissions reduction is the only game in town.

Our new situation is risky—but properly focused actions that shadow the high Arctic and remove CO_2 can greatly improve our chances. Doing something big should bring hope to the public during the 10-15 years it will take to begin reducing extreme weather.

There are effective actions we can still take to repel the extreme weather invasion, if we only get our act together in a hurry. Like war, it is risky—but properly focused actions can greatly improve our chances. *The trip to Hell is not a sure thing.*

Questions?

. . .

Q&A

"How many big emergency projects are we talking about?"

A: Depends on who you ask. If *emergency* is the operative term, I'd list two:

    1. *Carbon dioxide removal* to cool things off and reverse ocean acidification. It should be complete before 2040 because of the extreme weather surges, and needs new technology designed via a Manhattan Project 2.0.

    2. *Shading for the high Arctic* within five years—not the more difficult entire-globe version of solar radiation management, but only one percent of it, an Arctic Halo that tries to tame the extreme weather and thus protect the other projects for a few decades.

Both now require inventions, design, prototypes to test, and then fast implementation within five years. If we knew a meteor would strike the earth in twenty years, it would be a similar emergency. *This is not your grandfather's global warming anymore.*

. . .

The two projects which are *urgent* may instead require a lot of new infrastructure and social engineering, but they are not tech fixes:

    3. *Relocation* to uphill of coastal plains; that needs to be complete well before 2050. Some major cities might instead become an island surrounded by sea walls and pumping stations.

    4. Preparing for *inland extreme weather refugees*. Extreme weather is arriving much sooner than the sea level rise, and it is everywhere, not merely coastal plains.

Yes, they are expensive but there are no alternatives except for giving up, which is what further delay has now become.

...

In the third rank, I would place:

    5. Recreating the *Pharaoh's Seven-year Grain Storage Plan*, in case of crop failures from extreme weather.

    6. We need to put the economists to work designing a wartime economy, what various sides did during WW2. Events that would destabilize the global economy need to be identified in advance, and some stabilizing mechanisms created to get us through a rough period.

This is just my short list. There will be more.

...

Just remember the default answers, should we fail to act soon enough: *Life becomes cheap, no longer precious.* Many will tune out the continuing stream of bad news and just hunker down, thinking that most of their children are not going to survive very long either. Many will be angry about it and act irresponsibly.

    That state of affairs can still be avoided. Whenever climate pessimism comes up, I remind people that, in dealing with Phase Two of climate disease, *we really haven't tried yet* (emissions reduction doesn't count except as prevention)—and certainly not with the kind of national mobilization to get the job done that some of us can recall from World War Two. I can't believe that we will give up without even trying. Yet nearly all of our public discussion these days fails to mention any of the above six projects.

...

The prognosis for sea level rise is bad enough. But the biggest threat is severely disrupted agriculture from the heatwaves, floods, and windstorms. Let me briefly spell out what that means, in terms I would ordinarily reserve for an adult audience. Skip the next three paragraphs if you must; they are a brief description of hell on earth.

> When other countries get hit at the same time, importing food doesn't work anymore. As winter approaches, there are food riots. Famine leads to raiding neighbors and then resource wars, refugees on the move, genocides, and epidemics (when people drink from ponds and lakes that were contaminated by last week's group of refugees rinsing diapers).

Such multi-impact societal collapses happened in ancient agricultural times, and often enough for the combination to get a name: *apocalypse*. I just mentioned all four horsemen[60] and their underlying actions.

The result would be a human population crash—one in ten might survive—with the next generation responding only to authoritarian rule and often missing out on education—yet having to dig themselves out of a deep hole over many centuries while re-inventing civilization. This would be slow because each of the surviving pockets of humanity would hate their neighbors for events during the downsizing.

Our best chance of avoiding this fate, it now appears, is to repair climate very quickly.

Our situation is not hopeless, as some commentators are beginning to suggest—ones who have been led to believe that emissions reduction is the only game in town. Our new situation is risky—but properly focused actions that shadow the high Arctic and quickly remove CO_2 can greatly improve our chances.

"Sea level rise? Has that train left the station?"

A: On the time scale of 2050, I'm afraid it has. Even if we reverse the overheating of the land surface by 2040 with CO_2 removal, the oceans will take much longer to cool. Most of sea level rise is from the thermal expansion of the warmer surface layer. I expect that the sea level threat will be with us for the rest of the 21st century. Given that we had fifty years of warning, that's self-inflicted harm.

Reversing thermal expansion can be done but reversing the second source of sea level rise, the melting of land-based ice, means building ice sheets up again, taking water out of circulation. And even that won't restore things; the bottom of some glaciers has become unstuck from the frozen ground below, thanks to meltwater draining down cracks uphill. So, there is now rotten, easily fractured ice in many places, and it take forever to replace. Even if we can cool enough to add new ice layers on top, the rate at which the old and new ice slides downhill may not slow very much.

The resettlement from the vulnerable coastlines, at least in the areas with low coastal plains, needs to start immediately if we are to avoid refugee and economic crises in the next several decades.

While this requires social engineering and economic emergency measures, of the sort not seen since WW2 military planning, it doesn't need new technology to be invented and deployed, not in the way that cooling us off does. It certainly will be a boom time for the construction industry if the economists can keep the economy from crashing.

"Say more about that chain of causes."

A: We often distinguish between an ultimate cause, such as burning fossil fuels, and the proximate cause of the extreme weather—say, those jet stream kinks. Names for intermediate causes include secondary and tertiary.

As is common in medical science, interventions can target any level—say, where a halo of bright haze to shade the high Arctic might be effective against the Arctic's vicious cycle and thus much of the new extreme weather, even though it would not cool the rest of the world very much.

Hereafter is where the abridged softcover editions differ from the full-length Kindle edition, as 80 *printed* pages have been saved by posting the rest of the **Q&A** and all the **Endnotes** on the web as a free PDF at *CO2Foundation.org*. They are essential reading for those interested in how CO2 removal could be done, as well as the 1% Arctic Halo.

www.ingramcontent.com/pod-product-compliance
Lightning Source LLC
Chambersburg PA
CBHW040314220526
45473CB00009B/2428